Springer Polar Sciences

Series editor
James Ford, Priestley International Centre for Climate, University of Leeds,
Leeds, UK

Springer Polar Sciences

Springer Polar Sciences is an interdisciplinary book series that is dedicated to research on the Arctic and sub-Arctic regions and Antarctic. The series aims to present a broad platform that will include both the sciences and humanities and to facilitate exchange of knowledge between the various polar science communities.

Topics and perspectives will be broad and will include but not be limited to climate change impacts, environmental change, polar ecology, governance, health, economics, indigenous populations, tourism and resource extraction activities.

Books published in the series will have ready appeal to scientists, students and policy makers.

More information about this series at http://www.springer.com/series/15180

Niels Vestergaard • Brooks A. Kaiser
Linda Fernandez • Joan Nymand Larsen
Editors

Arctic Marine Resource Governance and Development

Springer

Editors
Niels Vestergaard
Department of Sociology, Environmental
and Business Economics
University of Southern Denmark
Esbjerg, Denmark

Brooks A. Kaiser
Department of Sociology, Environmental
and Business Economics
University of Southern Denmark
Esbjerg, Denmark

Linda Fernandez
Department of Economics
Center for Environmental Studies
Virginia Commonwealth University
Richmond, VA, USA

Joan Nymand Larsen
Stefansson Arctic Institute
University of Akureyri
Akureyri, Iceland

ISSN 2510-0475   ISSN 2510-0483   (electronic)
Springer Polar Sciences
ISBN 978-3-319-88420-2   ISBN 978-3-319-67365-3   (eBook)
https://doi.org/10.1007/978-3-319-67365-3

Printed on acid-free paper

This Springer imprint is published by Springer Nature
The registered company is Springer International Publishing AG
The registered company address is: Gewerbestrasse 11, 6330 Cham, Switzerland

Preface

The rates of climate and environmental change in the Arctic region greatly exceed global averages and are accelerating in response to biophysical feedbacks connected with human economic activities. How can societies address the complexities of resource development in the Arctic and also create more effective governance systems that promote social justice and the rights of indigenous peoples? This important volume, the outcome of a conference on Arctic Marine Resource Governance in Reykjavik, Iceland, in October 2015, addresses opportunities for implementation of creative resource management systems and new institutional arrangements and technology for sustainable development.

As sea ice declines, the emerging frontier of a new Arctic Ocean presents serious challenges to existing governance models, international regimes, and agreements. The editors have assembled a diverse group of leading scholars who use disciplinary and interdisciplinary approaches to focus on issues in the broad and overlapping categories of Arctic governance, fisheries, and technology and development.

Reaching across important issues of energy resource extraction economics and the infrastructure demands for shipping, the authors provide a much-needed dialogue on fisheries regulation, management, and protection of the Arctic marine environment.

They critique the use of a trans-boundary governance capacity framework to evaluate key relationships among the Arctic states and more recent actors such as China. A discussion of the rise of Asian state interests in the Arctic; the complex histories of political interactions between the USA, Canada, and Russia; and the distinct interests of indigenous peoples' organizations of the Arctic is a critical bridge linking the individual chapters. The authors are convincing in making a case that the Arctic Council must be flexible and adaptive across multiple scales and uses if it is to remain the key governance forum for the Arctic region.

The core of the book covers the challenge of reaching sustainable management of living resources. The economic dimensions of fisheries (and resource) governance differ across the scale of indigenous communities to industrial sectors. Sustainable resource development for all must keep pace with rapid climate change, shifting geopolitical interests in the Arctic, and lagging investments in critical infrastructure such as ports.

As colead scholars for the Fulbright Arctic Initiative, we had the exciting opportunity to work with interdisciplinary teams of scholars to create innovative applied research on energy, water, health, and infrastructure policy challenges. This volume meets these same objectives. The authors have created a much-needed, accessible, thoughtful, and forward-looking perspective on the future of resource development and governance in a rapidly changing Arctic.

Institute of Arctic Studies   Ross A. Virginia
Dickey Center for International Understanding
Dartmouth College, Hanover, NH, 03755, USA

Woodrow Wilson Center for International Scholars   Michael Sfraga
Washington, DC, 20004, USA

International Arctic Research Center
University of Alaska Fairbanks
Fairbanks, AK, 99775, USA

Contents

Introduction

"The ecosystem changes underway in the Arctic region are expected to have significant impacts on living resources in both the short and long run, and current actions and policies adopted over such resource governance will have serious and ultimately irreversible consequences in the near and long terms." These were the opening words for the call to the Arctic Marine Resource Governance and Policy conference held in Reykjavik, Iceland, in October 2015 with various scholars and practitioners. The conference was supported through grants from Nordregio, the Carlsberg Foundation, and the Nordic Council of Ministers. The themes of the conference were:

1. Global management and institutions for Arctic marine resources
2. Resource stewards and users: local and indigenous co-management
3. Governance gaps in Arctic marine resource management
4. Multi-scale, ecosystem-based, Arctic marine resource management

These themes are all based on the observation that the global interest in economic development in the Arctic has been growing rapidly. This has been driven by both supply and demand shifts for the resources and amenities produced in the Arctic. On the cost side, there is a perception that climate change-driven impacts in the Arctic will reduce ice cover (both sea ice and land-fast ice) in the area and therefore also access costs and the broader costs of doing business in five particular industries: shipping, oil and gas exploitation, mineral resource extraction, fisheries, and tourism. Simultaneously, the global demand for these resources and amenities is rising as both population and wealth increase. However, the complexities of resource development in the Arctic (e.g., natural systems poorly understood, national versus international interests, and the rights of indigenous peoples) point to the need for addressing this by developing new institutional arrangements and creative solutions. The development of new governance structures must include the fact that ecosystems will be changing in the next many decades. For example, Arctic marine resources – whether mobile or stationary – may straddle various political and geography boundaries at regional and international scales, and the governance structure has to include this as a core assumption. The second feature of the governance structure is sustainability which will require (1) a balanced utilization between local (indigenous people) and global demand and (2) a balanced utilization between the maritime sectors to be able to capture the total economic value of the ecosystem services from the Arctic marine resources. With climate change and a potential opening up of the Arctic region as a new frontier, the business opportunities are many. The chapters in this book volume address the research questions just outlined in detail.

Bertelsen, in his chapter "The International Political Systemic Context of Arctic Marine Resource Governance," seeks to identify and characterize the international systemic framework of marine resource governance in the Arctic. He looks closely at ecosystem changes in the Arctic marine environment together with the adaptation of resource governance that comes along, and he also analyzes in the same context developments in the international political, economic, security, and legal system. The study purports to contribute to forming policy and research agendas in the Arctic and, more broadly, to a changing international system by means of adapting global governance. There are two main international systemic developments that the author considers of key relevance to the Arctic and analyzes thoroughly throughout his article. These are (1) the strategic competition between the USA and Asia (China mostly) and (2) the international position of "post-Soviet" Russia in the Arctic, which has shifted along with the Ukraine crisis. More specifically, he looks at East and West power transitions among Arctic and non-Arctic states and uses examples of international history in order to draw useful intuition of relevance to the Arctic. He describes in detail how China's and other Asian countries' interest in the Arctic has developed in recent years, and he discusses the conflict between Western countries and Russia on the Ukraine crisis along with the potential of a spillover onto Arctic governance issues. The discussion of Mearsheimer's views on likely military conflict between China and the USA provides a useful context of *The Tragedy of Great Power Politics*, to wit: "the stopping power of water," by which he suggests Asia, the Americas, and Europe will necessarily seek primacy only in their own continental areas (Callahan, 2015). Thus, the marine environment is a physical presence to reckon with for any power play. The author expands his analysis and looks into the role of Arctic transnational knowledge networks concerning Arctic marine resource governance using "science diplomacy" and "epistemic communities" as his main tools. "Science diplomacy" refers to the way in which transnational scientific collaboration, among countries with possible conflicts, helps build up trust. The author considers this particularly relevant for the case of power transition to China, which has created some distrust. Epistemic communities are transnational expert networks with an influence on policy-making, which apply in Arctic marine resource governance through the PAME working group of the Arctic Council, the University of the Arctic, research organizations such as IASC and IASSA, some large Arctic conferences, and other transnational Arctic research ties.

In "Reshaping Energy Governance in the Arctic? Assessing the Implications of LNG for European Shipping Companies," Holmes, McCauley, and Hanley review and analyze the transition away from heavy oil-based fuels used for maritime transportation toward the utilization of liquefied natural gas (LNG) as a replacement. As environmental regulations are prompting this transition toward more environmental alternatives, LNG is emerging as a likely and viable option that comes along with a number of socioeconomic costs and benefits. The authors tease out the key governance challenges involved in managing the transition toward LNG in shipping and use the Arctic region as a case study that reflects the challenges and opportunities for the maritime industry. Past energy transitions, along with governance mechanisms, are highlighted since they form a useful baseline for policy-making in future energy transition such as toward LNG. The analysis centers around three main governance challenges that arise from the upcoming transition and shipping. The first challenge lies in the governance within the shipping companies and the ways in which gas-based fuel can be developed. The second challenge refers to the expected increase in energy exploitation in the Arctic facilitated by the advances in hydrocarbon delivery as the icy barriers retreat and LNG features a less polluting future. The third challenge is the inclusion of shipping companies as stakeholders in the region, actively involved in policy decision-making, instead of simply being regulated by other authorities. The framework of the chapter suggests that the institutional changes needed to address those challenges can lead to future energy transitions in shipping and beyond, so that barriers and limits to economic growth can be removed.

Rayfuse's chapter, "Regulating Fisheries in the Central Arctic Ocean (CAO)," offers a well-informed and thoughtful analysis with relevant review of milestones in the momentum within the Arctic policy community over the Central Arctic Ocean. Understanding the key issues associated with the Oslo Declaration and the negotiation process of the Arctic 5+5 (Canada, Denmark, Norway, Russia, USA and China, European Union, Iceland, Japan, S. Korea) as well as the state of the scientific assessments regarding potential fish stocks in the Central Arctic Ocean Declaration (2015) may prevent "unregulated" commercial fishing. The existing agreement affects vessels under the jurisdiction of one of the coastal states, but also sets the broader global agenda. This agenda begins with the Arctic 5+5 negotiations. Such prevention, even if only among the Arctic Five, is not a negligible result, especially regarding possible operations by Russian fishing vessels, which could easily operate in areas of the CAO outside the jurisdiction of either the North East Atlantic Fisheries Commission (NEAFC) or the Joint Norwegian-Russian Commission. Regarding the Arctic 5+5 negotiations, the situation is evolving so quickly that the analysis in the chapter may well be out of date by the time this chapter is published. The late 2016 meeting in Torshavn (Faroe Islands) produced an expectation that some legally binding agreement would be reached in the near future. The author argues persuasively that this is unlikely to be in the form of a regional fisheries management organization (RFMO) and there is a legitimate question regarding whether the Arctic 5+5 format covers all the actors that might take an interest in potential CAO fisheries.

Snook, Cunsolo, and Morris in their chapter "A Half Century in the Making: Governing Commercial Fisheries Through Indigenous Marine Co-management and the Torngat Joint Fisheries Board" discuss indigenous co-management in the Labrador Inuit settlement region of Nunatsiavut. For the purposes of their analysis, they use as a case study the Labrador Inuit Land Claims Agreement and the Torngat Joint Fisheries Board (TJFB), which has contributed significantly in advising on commercial fishing and more broadly in integrating science and traditional knowledge as well as in policy planning and decision-making. The authors offer an overview of the status and challenges of land claims-based indigenous fisheries co-management in Canada and zoom into the Inuit land claims in Nunatsiavut. The TJFB, the first land claims-based co-management, evolved as a result of the long-term efforts for the establishment of an Inuit self-governing land claims settlement area in Labrador, along with ratification of the agreement which led to self-government for the Inuit of Labrador. The ways in which TJFB has impacted Arctic marine governance are being exemplified through three cases of economically vital commercial fisheries for Nunatsiavut: the Northern shrimp, the snow crab, and the Arctic char. In their review of the fish management policies of those species, the authors highlight the marginalization and exclusion of the Labrador Inuit in decision-making over the years and reflect on how indigenous co-management boards such as TJFB have helped build up healthier communities and ecosystems. Meanwhile, they also stress the need for a shift away from legal interpretation of the land claims documents and instead call for more emphasis in the "spirit and intent of these documents as living and breathing entities," implying that they should be used as a minimum baseline for future decision-making that will be focused on the support of the health of the environment and the people. The recently signed United Nations Declaration on the Rights of Indigenous Peoples is seen as a great opportunity for promoting indigenous rights, sovereignty, and reconciliation using the already-existing frameworks, which in an Arctic marine resource context can be achieved by strengthening co-management boards and including them in global decision-making.

Vestergaard in his chapter "Scenario Analysis for Arctic Marine Resource Policy" states that future changes in Arctic marine ecosystems will depend as much on global climate change as on our ability to regulate and manage exploitation pressure at sustainable levels. There is a lack of integrated, cross-sectoral, ecosystem-based analysis of the Arctic marine management. The analysis is on both the choices for implementing regulatory tools and how they will affect the many ecosystem-dependent values derived from them. In this chapter, likely changes in future Arctic fisheries based on a scenario building approach are speculated. The underlying climate changes to ecosystems and their likely impacts in the Arctic are also among the drivers. Other drivers can be identified such as the sectoral development of important marine sectors (fishing, shipping, mining, etc.) and governance structure development. By selecting two main drivers, it is possible to map four scenarios to be further analyzed. The development in each of these driving forces' dimensions is uncertain, and central in the analysis is risk and uncertainty. The results indicate that the future might involve relatively large future climate changes in the ecosystem and hence fish stocks and also that the economic outcome of fisheries depends critically upon the ability to adjust the regulatory regime to capture the values of the ecosystem services.

Kaiser and Parchomenko, in their chapter "Long Run Transitions in Resource-Based Inuit Communities," explore transitions of socio-ecological systems in Inuit communities in the Arctic and analyze the driving forces for those transitions. Natural resources, humans (viewed as harvesters), and resource user-managers holding specific human capital in the Inuit communities are in the epicenter of the analysis. The multi-trophic model sketched in this chapter sheds light on the ways in which early trade with non-Inuit communities affected long-run opportunities for Inuit. Using theories and tools from new institutional economics and resource economics, the authors look at the institutional changes across time as well as at the way in which governance has been evolving. Human capital or traditional ecological knowledge (TEK) is seen as top trophic level that has the potential of increasing the efficacy of resource harvesting (along with labor productivity and pressure on the resource) or increasing the base resource's productivity, which may have a series of different implications. For the purposes of exploring the co-evolution of governance structures and resource pressures, the authors look carefully into resource value changes, harvesting and governance costs for common property resources, and enforcement costs for wealth accumulation as well as trade. Historical evidence is linked to the different stages of Inuit economic development. The nuance of the chapter is that it goes beyond analysis of population dynamics and examines migration patterns, trade, and technological progress in tandem with dynamic forces in the resource-based Inuit economy. The multi-trophic model developed to explain the transitions in Inuit communities also includes an intertemporal dimension, broken down into value throughout time from consumption, from human capital formation and from trade, as well as current costs from harvest, dispersion, and enforcement of nonconsumption. The socio-ecological framework, for the case of the Inuit, centers on governance issues relevant to long-run sustainable development from an ecosystem base. The authors use two examples that help build a basic understanding of the limited development of population-conservation-based governance in Inuit communities: the trade introduction due to Russian and European fur interest and the whaling and walrusing trade example in which TEK, as a means of technology, played a major role. Overall, the case of the Inuit communities provides useful insights on the roles of governance in economic growth through trade, (limited) resource governance, ecological knowledge, and exploitation, and it may also provide pragmatic suggestions for facing the increasing resource scarcities in our planet.

The chapter "Ballast Water and Invasive Species in the Arctic," by Holbech and Pedersen, addresses the risks of nonindigenous invasive species introductions in the Arctic via ballast water. The authors discuss the invasive species risks posed by ballast water management in the context of the Ballast Water Management Convention, which is expected to enter into force in September 2017, as well as of the US Coast Guard Ballast Water Discharge Standard, which is already in force since 2012. They aim at providing an overview of the international and regional ballast water regulations with an impact in the Arctic region. They briefly outline the pathway toward the ratification of the convention and discuss the details of the ballast water management system itself, which comes down to a two-phased strategy. The phases are the regulation for the exchange and the performance standard. Regional ballast water regulations for the USA, Canada, and HELCOM/OSPAR member parties are also discussed throughout the chapter and analyzed in terms of their alignment with the IMO regulations. The chapter also sheds light on the way in which climate change affects the risk of invasive species introduction in the Arctic, pointing out the expected increase in shipping as the main vector of invasions either via ballast water or hull fouling. A comprehensive review of the specific challenges of shipping activity in the Arctic region is provided throughout the chapter, and special emphasis is put on the drivers of those challenges, such as the upcoming industrial development, touristic activity, etc. The vulnerability of the Arctic marine ecosystem is examined in terms of species richness and functional complementarity within trophic levels, which helps build a baseline for understanding the resilience and stability of the ecosystem. Temperature and salinity fluctuations are considered as key parameters for changes in the living conditions of Arctic marine organisms such as plankton. Among the challenges of ballast water management in the Arctic, the costs of installation and maintenance seem to play a discernible role. The authors look closely at the efficiency of the ballast water regulations in place for the prevention of invasions in the Arctic, by reviewing cases where exemptions apply and ballast water exchange can fail or cannot prevent the transfer of very small organisms. Attention is also drawn at the risks of hull fouling as a vector of invasive species introductions in the Arctic.

Pahl and Kaiser, in their chapter "Arctic Port Development," explore the potential for development of Arctic ports from both a logistics and an infrastructural point of view that also accounts for local concerns – an understudied issue in the existing literature for shipping's impact in the Arctic. Interests in Arctic infrastructure differ across space, and they have also been shifting in recent years. Indicatively, the drop in oil and gas prices has reduced the great interest in offshore oil and gas ventures to varying degrees across the Arctic, while in the Western Arctic, there has been a greater focus on infrastructure needed for the support of local community needs. The similarities and differences in the needs and potential of the three Arctic routes, the Northwest Passage (NWP), the Northern Sea Route (NSR), and the Transpolar Route (TPR), are thoroughly discussed throughout the chapter in the context of port infrastructure investment so as to help articulate the concerns for governance and coordination in Arctic development. The chapter seeks to identify how Arctic ports can respond to those challenges and concerns as well as those economic and environmental factors that will determine their future planning. Meanwhile, the authors look at how port logistics demand and the required infrastructure affect the development of Northern communities and their resources. In order to identify the actions needed that will boost positive and reduce negative influences of economic development in the North, the authors look closely into the demand and supply for port development that is required for safe and secure shipping in the Arctic. They provide a detailed review of all existing Arctic port infrastructure (in the USA, Canada, Iceland, Norway, Russian Federation, Greenland, and Faeroe Islands) along with the future strategies for their development. They also delve into the economic attractiveness of the three new Arctic routes and discuss their predicted future developments. They look at the complementarity of the relationships between the three routes and point out that it is more likely for the TPR to be a substitute for shore-based Arctic shipping since it is expected to help avoid economic and environmental challenges usually found in shore-based routes. Examples of bilateral and multilateral cooperation among Arctic countries and institutional settings such as the Arctic Council help build an understanding of the existing status of coordination and governance as well as of future needs. The authors duly stress the contribution of the Arctic Council as a platform for cooperation, which they demonstrate through the two recently signed legally binding agreements for safety and security and environmental risks. Both agreements are vital given no one country has ample resources on its own for the potential risks of more growth related to port development. The role of military use in port development is also considered in the context of coordination and governance.

Kaiser, Pahl, and Horbel's companion chapter "Arctic Ports: Local Community Development Issues" expands upon the port descriptions and analysis in the previous chapter. The article provides a description of a variety of Arctic communities with potential for developing resource extraction, tourism, and fisheries that transportation and port development support. Their evaluation of historical rail development along with rail and port networks in North America highlights issues that are relevant for Arctic marine ports and the communities they affect. These include the importance of network externalities, challenges of high fixed cost infrastructure development, and shifts in population and economic activities that may accompany port developments. Characterizing options such as four possible states of tourism among different communities within the Arctic include examples of actual resulting impacts on the communities.

The book volume is concluded with a chapter where the potential and future role of the Arctic Council is investigated. Could the Arctic Council act as the core developer of a governance structure securing balanced and sustainable economic development? The answer to this question is that the Arctic Council has the potential but needs to cover more areas and include more states and also to develop decision-making fora so cooperative solutions are in fact implemented.

Reference

Callahan, B. (2015). 2015-07-08-last update, *Mearsheimer vs. Nye on the Rise of China* [Homepage of The Diplomat] [Online]. Available: http://thediplomat.com/2015/07/mearsheimer-vs-nye-on-the-rise-of-china/ [2017, 03-21].

Part I
Governance

The International Political Systemic Context of Arctic Marine Resource Governance

Rasmus Gjedssø Bertelsen

Abstract The Arctic has been an integrated part of the international system for centuries, and systemic developments have deeply influenced the region and its communities. Central Arctic Ocean marine resource governance is in the nexus of climate change and international systemic developments. The international systemic context for the Arctic is: The rise of China and emerging Asian economies driving gradual power transition from Western to Eastern states. Struggles continue over the domestic order and international position of post-Soviet Russia, where either side considers whether to escalate the Ukraine crisis horizontally to the Arctic. The USA and China interact concerning governing Arctic marine resources as Arctic Ocean coastal state/status quo power and fishing nation/rising power. Russia and the West choose not to escalate the Ukraine crisis horizontally into Arctic marine resource management. Co-creating of knowledge and epistemic communities are important for Arctic status quo and rising Asian countries to manage power transition in the Arctic and for Russia and the West to continue Arctic cooperation despite political crisis elsewhere.

Keywords International system • USA • China • Russia • Power transition • Post-Soviet • Ukraine-crisis • Status quo power • Rising power • Globalization

R.G. Bertelsen (✉)
UiT-The Arctic University of Norway, Tromsø, Norway
e-mail: rasmus.bertelsen@uit.no

© Springer International Publishing AG 2018
N. Vestergaard et al. (eds.), *Arctic Marine Resource Governance and Development*,
Springer Polar Sciences, https://doi.org/10.1007/978-3-319-67365-3_1

Introduction: Seeing the Arctic as Part of an International System

Arctic marine resource governance takes place at the interface of global natural and social systems (Bertelsen 2014). This book looks at the major ecosystem changes under way in the Arctic and their effects on living resources and subsequently needs for adaptation of marine resource governance. These ecosystem changes are especially driven by climate change, which is a clear example of a complex global socio-techno-environmental system. The ecosystem changes and responses in terms of governance of Arctic marine resources happen within a context of the international political, economic, security and legal system.

The aim of this chapter is to place these Arctic marine ecosystem changes and adaptation of resource governance within the context of the parallel developments of the international political, economic and security system ("the international system"). The chapter seeks to set out this international systemic framework for the other chapters on Arctic marine resource governance from biological, economic, legal perspectives and at different scales. For instance, the chapter by Rosemary Rayfuse deals in detail with how the Arctic Ocean coastal states seek to adapt governance of Central Arctic Ocean marine resources to climate change and international systemic change through international law with the Oslo Declaration and its Broader Process. The chapter here focuses on the international systemic framework and how Arctic and outside actors use transnational knowledge networks and epistemic communities to adapt Arctic governance, including of marine resources, to international systemic change. The chapter seeks to contribute both to Arctic studies and to general policy and research agendas on adapting global governance to a changing international system.

An example of Arctic marine research governance adapting to climate change and international systemic change is regulating fisheries in the Central Arctic Ocean (CAO). CAO becomes accessible because of climate change. Politically and legally, CAO are high seas open to all, but also the backyard of the status quo powers of the international system (USA, Russia (a status quo Arctic power), Canada, Kingdom of Denmark and Norway), while the rising powers are China and other Asian emerging economies. CAO therefore illustrates how the governance of Arctic marine resources takes places at the intersection of global environmental change and global political change. There is both legal regulation of this challenge, and there is a framework of transnational knowledge networks and epistemic communities where this problem is discussed and common understanding and solutions are sought.

There are two key international systemic developments of relevance to the Arctic, including marine resource governance. They are first the strategic competition between the USA and China in a context of gradual power transition from West to East with a relatively more affluent and powerful China and other Asian emerging economies. The other relevant international systemic development is the continuing struggle over the domestic socio-economic-political order and international position of post-Soviet Russia.

Lay observers of international politics may not intuitively think in terms of a system with interactive effects or unintended consequences. A systemic approach to international politics is useful as there are important systemic traits in terms of the system behaving in ways not desired by the units, interactive effects and unintended consequences – as ecological systems (Jervis 1997). The basic characteristics of the international system are that the main actors are states, who exist under anarchy (the absence of a higher authority) (Waltz 1979). Therefore, states watch each other with suspicion and may easily fall into security dilemmas, where one state seeks security in armaments, which provokes fear among other states, who arm in response, cancelling out the added security of the first state (Herz 1950). Even if not resorting to war, anarchy and fear of unbalanced relative gains can be a major impediment for states to collaborate to solve common problems and receive absolute gains.

A key property of the international system is the development over time of the relative power between the most powerful great powers, which is termed power transition. This relative power shifts with different growth rates and a new challenger power catches up and surpasses the existing leading status quo power. Such power transitions or attempts of power transition have often been exceptionally dangerous episodes in international history as illustrated by the Napoleonic wars, two world wars, or the Cold War (Organski 1968 [1958]). Based on the distribution of power among the most powerful great powers, the system can be described as multipolar (roughly before World War Two), bipolar (the Cold War with two super powers) or unipolar (the USA as sole superpower after the dissolution of the USSR), which has also shaped the Arctic and its communities.

A Brief Historical Background of the Arctic in the International System

In order to place Arctic marine resource governance in the international system today and draw lessons, it is useful to briefly outline the historical embeddedness of the Arctic in the international system concerning politics, economics and security – where natural resources have often been the connection. The place of the Arctic in these larger systems has often had profound impact on societies in the Arctic.[1] There has been an unfortunate tendency in popular and even academic writing on the Arctic to suggest a risk of conflict in the Arctic driven by competition over local natural resources made accessible by climate change (Borgerson 2008). Here a historical perspective makes it clear that conflict has mainly been driven by general international conflict at the systemic level. This is the case concerning traditional military national security and comprehensive human and environmental security (Gjørv et al. 2014).

---

[1]The interconnectedness and migrations of Arctic indigenous peoples with the outside is acknowlededged, but that is outside the scope of this paper and the competences of the author.

The Arctic has been a part of the Western economic system often for as long as Westerners have lived there based on extracting and exporting natural resources, both biological and mineral. North Atlantic waters off the coast of Norway, around Iceland and off the Atlantic coast of Canada are particularly rich fishing and whaling grounds. North Atlantic fisheries and stock fish have been a part of the European food system since the middle ages. This fish fed large parts of Catholic Europe especially during Lent, reflected today in the bacalao dish in Portugal and Spain. Svalbard was the site of intensive Dutch, British and other whaling since the 1600s and later coal mining. The Hudson Bay Company was founded in 1670 to bring North American furs to European markets and is one of the oldest existing companies of Western capitalism today. Arkhangelsk on the White Sea was the international trading port of Russian before Peter the Great founded St Petersburg in 1703, as is displayed in the Merchant Court museum in Arkhangelsk. Iron-ore mining in Northern Sweden has been a key part of Swedish industrial economy and formed a complex socio-technical megasystem since the late 1800s (Hansson 1998). In April 1940, German occupied Denmark (as stepping stone) and Norway to secure iron-ore shipments from Narvik. The role of Northern Swedish iron ore in German heavy industry during World War Two is just one example of the international strategic importance of and embeddedness of the Arctic and Arctic resources. The USSR greatly industrialized the Soviet Arctic to extract oil, gas, minerals and other natural resources. Alaska is one of the most important oil producing states of the USA, and waters off Alaska provide a large share of seafood consumed in the USA (Heininen and Southcott 2010).

The North Atlantic is the connection between Europe and North America. Maintaining or breaking this connection is key to European-North American great power politics. Western great power conflict has deeply affected the Arctic from the Seven Year's War (France's loss of Canada), the Napoleonic Wars (separation of Denmark and Norway), the Crimean War (Russian sale of Alaska to the USA), World War I (German unrestricted submarine warfare, Imperial Russian establishment of Romanov-on-Murman (Murmansk)) and World War II to the Cold War. Britain occupied the Faroe Islands on the 12th of April, which led to de facto Faroese independence during the war laying the ground for home rule from 1948, and Iceland on the 10th of May. Britain was relieved by the USA in Iceland on the 7th of July 1941, 5 months before the attack on Pearl Harbor. The North Atlantic and the Barents Sea again became key battle grounds as Germany tried to cut off Britain and also cut off the allied convoys to Murmansk. As part of Operation Barbarossa, Germany invaded the Kola Peninsula from Finnmark in Norway and Lapland in Finland seeking to conquer Murmansk and cut off the USSR. The Pacific Arctic was also very much a theatre of WWII. Alaska is an American staging ground towards the Asia-Pacific region in addition to California, and Japan invaded the Aleutian Islands, which became the Arctic battle fields of the USA.

The Arctic was exceptionally militarized during the Cold War with strategic nuclear weapons systems and early warning systems. The trajectory across the North Pole is the shortest between North America and Eurasia for long-range flying, intercontinental ballistic missiles, submarines and nuclear weapons. The Cold War ended in the Arctic with Mikhail Gorbachev's speech in Murmansk in 1987 calling for the Arctic as a zone of peace, research and environmental collaboration. The current Arctic with Circumpolar cooperation was only possible because of the end of the Cold War: the Rovaniemi Process of 1989, Arctic Environmental Protection Strategy of 1991, Norwegian-led Barents collaboration from 1993, Ottawa declaration of 1996 founding the Arctic Council, the International Arctic Social Sciences Association and the International Arctic Science Committee as well as indigenous peoples' collaboration.

The Arctic in Today's International System: Globalization and Power Transition

As set out above, the Arctic has been an integrated part of the international system for centuries, and peace and war in the Arctic has not been driven by local conflicts, but great power rivalry. This is also the case for the Arctic in today's international system, where designing fisheries governance in the CAO in the nexus of climate change and power transition is one example.

Today's international systemic context for the Arctic is marked by two inter-linked mega trends, globalization and power transition. This chapter focuses on power transition, but it is useful to mention globalization first since globalization makes today's power transition qualitatively different than previous power transitions. One definition of globalization of use here is the "compression of time and space", that societies affect each other faster and with more impact than ever before (Harvey 1989).

The Arctic has in recent decades been the topic of new attention from new quarters (which probably contributed to the mistaken idea that the Arctic was previously isolated from international economic, political and security dynamics). This new attention reflects how globalization and power transition affects the Arctic. Arctic climate change is an instance of globalization. It has strong local effects as the diminishing sea-ice making CAO fisheries governance a concerning. However, climate change in the Arctic is caused far south in centers of population and economic activity, with amplified effects in the Arctic, and strong feedback effects in mid-latitude regions through the danger of sea-level rise and effects on climate and weather.

Local Arctic climate change (as in the CAO) and the feedback effects are usually the scientific rationales for outside states to engage in Arctic research, including China. Former US environmental diplomat and now climate advocate and chair of Arctic21, Rafe Pomerance, says to raise American public awareness of Arctic climate change: "the fate of the Greenland ice sheet is the fate of Miami". Likewise, the Chinese economy and population is concentrated along the coast, which is low and vulnerable to sea-level rise. Agriculture and food security is naturally a major concern to China with its enormous population, so possible adverse weather effects on Chinese agriculture from Arctic climate change is of concern and scientific interest (Li and Bertelsen 2013).

Power Transition: The Return of China and Asia – Also to the Arctic

Power transition from Western states to Eastern states is largely driven by the "rise of China" based on its spectacular economic growth since the Open-Door policy of the late 1970s. In a longer historical perspective, one can speak of a "return of China" in light of China's historical relative economic weight in the world. In addition to China, a number of especially Asian emerging markets have grown significantly in the afterwar period, especially the other four Asian countries engaged in the Arctic, Japan, South Korea, Singapore and India (Nye 2011). Globalization today makes China and other Asian emerging markets a much more acutely felt force around the world, than even relatively larger Chinese and Asian economies did centuries ago.

The USA and China are in terms of economic size, military capabilities, etc., increasingly each other's peer competitor in a bipolar international system, where the next level of great powers, including Russia, are distanced in terms of economy, military, etc.[2] The strategic competition and bipolarity between China and the USA with its allies and partners around the world affect regions, institutions and societies– including the Arctic. This Sino-American strategic competition is increasingly a context for Arctic marine resource governance, both concerning research and policy.

Arctic governance decisions are increasingly a question for the USA and the other Arctic states, who are all status quo powers in the international system, how to manage the ever rising outside interest in the Arctic from especially China and other advanced Asian economies. This was the question concerning granting regular observer-status to China, India, Japan, Singapore and South Korea to the Arctic Council in 2013. It is equally the fundamental political question behind how to govern CAO fisheries (although the Ilulissat Declaration of 2008 had also created divisions between the eight Arctic states themselves). China, India, Japan, Singapore and South Korea gaining regular observer-status in the Arctic Council are at the forefront of the rise or return of Asia. However, China being the strategic peer competitor of the USA stands out from the rest who are politically aligned with the USA and the West.

Managing the current and future Sino-US relations is a key policy and research question, which is structurally determined and has applied to power transitions throughout history. This structural question also applies to strategic choices in the Arctic. The USA and the other Arctic status quo powers must decide whether they will seek to exclude and suppress China as the rising power or try to integrate, accommodate and socialize China into the existing Arctic system. China as the rising power must decide whether to challenge the status quo or work and grow

---

[2]Keynote address by Professor Øystein Tunsjø, Norwegian Institute for Defence Studies, at NORASIA conference, 13th of January 2017.

within the institutions and organizations shaped by the status quo power. The two positions are clearly set out by the International Relations theorists, the offensive realist John Mearsheimer and the liberal institutionalist Joseph Nye (Callahan 2015).

Applying Mearsheimer's analysis (Mearsheimer 2014) and its implications to the Arctic would suggest that the Arctic states should strive to keep China and other rising powers out of the Arctic to preserve their own power in the region. China should be excluded from the Arctic Council, from investing in natural resources, engaging in scientific research, etc. At the Arctic Council ministerial meeting in Kiruna in May 2013, China would not have got observer status. Concerning Arctic marine resources and the case of the CAO, Arctic coastal states should strive to exclude China in institutions and practical fishing activities.

The Nye approach to China (Nye 2006) in the Arctic would be to integrate China in the institutions in order to socialize China and make China a responsible stakeholder in a stable Arctic. Concurrently the Arctic status quo states would seek to hedge against China violating this integration into existing institutions. It seems quite clear that the Arctic states are generally following a Nye strategy. China was integrated into the Arctic Council structure as observer state in May 2013 (together with the other four Asian states, but they are more or less allied of the West, so not challengers). China is increasingly participating in natural resource investments in the Arctic, especially in Russia (the Yamal LNG project). China is become a significant participant in Arctic research across disciplines and building transnational knowledge networks with the Arctic states, including the question of Arctic marine resources. The Broader Process of the Oslo Declaration for regulating CAO fisheries also serves as a framework for integrating China.

The power transition in relative economic, political, military, scientific, technological and cultural power from Western states to Eastern states and from state to non-state actors is also clear concerning the Arctic and explains much of the new attention to the Arctic from new quarters. For some centuries, the international system has been unquestionably Western-dominated. Western countries colonized the rest of the world and dominated it politically, economically, militarily, scientifically, technologically, linguistically, culturally, etc. We have become so accustomed to, for instance, interfering in or studying faraway parts of the world, that we in the West do not think twice about it. On the contrary, we are profoundly surprised when Asia starts to take a political, economic, scientific, etc., interest in our backyard, the Arctic. Bertelsen had an exchange with Teemu Polosaari at the Transarctic Agenda/Northern Research Forum 2015 conference in Reykjavik, where Polosaari expressed surprise that Bangladeshi scholars would be interested in the Arctic, where Bertelsen pointed out, that we would never be surprised by Western research interest in Bangladesh or anywhere else.

The return of China and the other four Asian Arctic Council observer states reflects political, economic and scientific power transition. China as the world's largest population and one of the world's two largest economies sees itself as a natural stakeholder in global governance around the world, including the Arctic and Arctic marine resource governance. China's investment and production-led growth has also turned the country into an enormous buyer of energy and raw materials, which is sourced globally. The Arctic is one possible – out of many – sources of energy (compare post-Ukraine gas-contracts with Russia) and raw materials. The Arctic is an expensive source of energy and raw materials, but it offers other advantages in terms of political stability unlike, most clearly, the Middle East (Li and Bertelsen 2013). China is also sourcing protein around the world, where Arctic fisheries are one possible source.

Post-Soviet Russia in the International System and the Arctic

The other international systemic development affecting the Arctic is as mentioned above the ongoing struggle over Russia's domestic socio-economic-political order and place in the world. The post-Soviet Russian struggle came to the fore with the Ukraine crisis since 2014. This process is the aftermath of the demise of the USSR and the end of Cold War bipolarity leading to temporary American unipolarity, which was a tectonic international shift. In light of Russia being by far the largest Arctic country in terms of territory, population, infrastructure, etc., the Arctic is also potentially deeply affected by this international systemic development. Russia has the longest Arctic coast line, which makes it a key state concerning Arctic shipping and fisheries governance. The question here is whether Russia, the USA and other Western states decide to escalate conflicts outside the Arctic horizontally to the Arctic or not in areas of marine resource governance or elsewhere.

The ongoing struggle over the place of post-Soviet Russia in the international system and the nature of Russian politics and society went off the rails in Ukraine in 2014 and threatens to spill into the Arctic through horizontal escalation. The dissolution of the USSR removed one of the two superpowers with a global military, ideological and economic reach. Post-Soviet Russia was for at least a decade a much-reduced regional power, which lost significant terrain by losing its Central and Eastern European satellite states and Soviet republics. Russia was retreating until 2008, when it fought a short war with Georgia to end its NATO ambitions. The European Union and NATO expanded eastwards invited by newly democratic countries. However, the threat of an EU agreement with Ukraine in 2014 was clearly too close to home for Russia, which first lost its client president and then intervened militarily annexing Crimea and covertly in Eastern Ukraine. These interventions are naturally unacceptable to the West and have caused a deep Russian-West crisis since (Mearsheimer 2014).

USA, Russia, the EU, Britain, France, Germany have myriad interfaces, and policy is much a question of choosing when to take conflict in one area into another area or vice versa. Russia has invaded its neighbor Ukraine; however, this action has led to much misguided alarmism concerning Russian expansionism in the Arctic, which is another example of the pitfalls of faulty analysis of the Arctic in international politics and security. Russia has a very specific strategic problem in

Ukraine, which it has no other means to solve than military force (Russia's great weakness in soft power reduces its options to hard power). However, Russia has no such problems in the Arctic, and therefore Russia has no interest in escalating the Ukraine crisis horizontally to the Arctic.

The West must find areas where it can hurt the political economy underpinning Russian power without risking military escalation. Russia has historically been and continues to be a natural resource-based economy, where oil and gas revenues (including from the Russian Arctic) play a large role. Western financial and technological sanctions of Russian Arctic offshore oil and gas developments are therefore a sophisticated way of threatening the future development of the Russian economy as punishment for the invasion of Ukraine.

Russia has made a point of continuing Arctic dialogue. Bertelsen was himself invited as a Danish academic with other Arctic and Asian Arctic Council observer state academics to the Arctic international high level meeting of the Russian Federation Security Council in Naryan-Mar in August 2014, where no Western diplomats were present, and Arkhangelsk in September 2015, when Western diplomats had returned. Western financial and technological sanctions of Russian Arctic energy projects made it necessary for Russia to pursue alternative partnerships. Russia entered into a number of large gas-contracts with China following the Ukraine crisis. Concerning Arctic marine resource governance, it is also clear that the USA and the three other NATO Arctic coastal states and Russia have been both able and willing during the Ukraine crisis to collaborate on the Oslo Declaration and the Broader Process. Russia has an obvious interest in contributing to Arctic cooperation in this field, and the USA and the three other NATO Arctic coastal states have decided not to escalate the Ukraine crisis horizontally into CAO.

Drawing Arctic and General Lessons on Arctic Marine Resource Governance in Today's International System

As mentioned above, this chapter seeks to both contribute to Arctic studies and general policy and research debates on global governance. The predominant international policy and research question today is how to manage power transition with the return of China in a globalized, compressed world. How to manage this power transition dominates policy and research questions from the highest levels of, for instance, US strategy to the Asia-Pacific region with the "pivot to Asia" with military and naval deployments and trade relationships as the Trans-Pacific Partnership. The election of Donald Trump as US president and uncertainties about US commitments to military alliances and trade agreements demonstrate how domestic politics can intervene in foreign policy, which is a key International Relations research topic, but beyond this chapter. A sub-question of the larger strategic question of how to respond to power transition in a globalized world concerns the Arctic and a further sub-question, Arctic marine resource governance.

Placing the topics of this book in that wider international systemic framework is therefore the aim here.

There has in recent years been discussion of China's and other Asian states' interests in the Arctic, where some voices have claimed a special interest of China in the Arctic, or that the Arctic should be particularly important for China. Here it is important to keep the global changes in mind. China's interests and actions in the Arctic reflect such general global trends rather than anything particular to the Arctic. The chapter will turn to look at the role of Arctic transnational knowledge networks and epistemic communities where the Arctic status quo powers and the rising Asian powers, in particular China, can co-create knowledge and build trust for adapting Arctic governance to power transition, including concerning marine resources.

Knowledge and Epistemic Communities for Managing Power Transition

When looking at how states manage power transition – or other complex processes – one should consider the role of information, perception, judgment and decision. One of the fundamental debates in social sciences and which has driven much research is between a standard rational actor model inspired by especially neo-classical economics and its critics in social and political psychology, behavioral economics and similar perspectives. The debate is between a parsimonious perspective that assumes rational actors with sufficient information and whom perceive change sufficiently accurately and quickly to act rationally in, for instance, such a dynamic system of power transition (Lau 2003). For the Arctic, it would mean that the USA and other Arctic states on one side and China on the other side understand what is happening and what is at stake. Processes through which knowledge is created and shared, perceptions formed and socialization takes place are not of importance.

The alternative perspective of social and political psychology (which is the focus here) or behavioral economics (which has a dynamic debate with the standard neo-classical perspective in economics) takes it starting point in psychology. According to psychology, humans simply do not satisfy the assumptions of the rational actor perspective. Humans do not have the cognitive and computational abilities, so they are forced to resort to a range of biases to function. Also previous, unrelated experiences have a significant influence on how later unrelated situations are perceived. So according to this perspective, what knowledge exists, who has access to it and takes part in creating it, how actors learn and are socialized is of importance (Sears 1975). According to this perspective, shared or separate knowledge, experiences, beliefs, etc., between Chinese, American and other Arctic decision-makers and stakeholders makes a difference how power transition plays out in the Arctic. And these factors of perception will make a difference in other settings of this power transition, which may make the Arctic lessons generally interesting for policy and research.

To look at the role of Arctic transnational knowledge networks concerning Arctic marine resource governance, we will focus on a few key concepts: science diplomacy and epistemic communities. The first concept is science (for) diplomacy, where scientific collaboration between countries with mutual mistrust and possible conflict builds parallel channels of communication and knowledge co-creation for building trust (The Royal Society, AAAS 2010). Power transition is often accompanied by deep distrust between status quo and rising powers, which is also clear concerning China in the Arctic, where science diplomacy can play a valuable trust-building role (Bertelsen et al. 2016). The second is epistemic communities, which are transnational expert networks, whose members agree on the nature of the problem and its solutions and have influence on policy (Haas 1992). Arctic marine resource governance is an example of a complex transnational problem, where expert communities play an important role.

Arctic Transnational Knowledge Networks and Epistemic Communities for Adapting Arctic Marine Resource Governance to Power Transition

The potential of science diplomacy for China to build trust among Arctic states and stakeholders through science diplomacy despite the inherent distrust of power transition has been analyzed and discussed (Bertelsen et al. 2016; Pan and Huntington 2016). There are a number of multilateral and bilateral Arctic transnational knowledge networks and possible epistemic communities, who play important roles for co-creating understanding of problems and solutions and building trust. These multilateral and bilateral networks will be outlined and their possible contributions to Arctic marine resource management discussed.

The key multilateral Arctic forum is naturally the Arctic Council with its six working groups and ad hoc task forces. This working group structure is some of the most important long-running epistemic communities concerning the Arctic. China and the other four Asian observer-states have access to participate in this working group work. Of the six working groups, the Protection of Arctic Marine Environment (PAME) is of course particularly important concerning Arctic marine resource governance. In line with what was explained above, Russia has clear interest in the good functioning of the Arctic Council and has kept the Ukraine crisis out of the work of the council.

There are three Arctic research organizations which are also central transnational knowledge networks and possible epistemic communities, the International Arctic Science Committee (IASC), the International Arctic Social Sciences Association (IASSA), and the University of the Arctic. Especially IASC has a very formalized structure as it primarily focuses on natural sciences, where logistical coordination of expensive research assets and resources play a very important role. University of the Arctic brings together close to 200 higher education institutions in the Arctic and outside with interest in the Arctic. The elaborate structure with transnational thematic networks and exchange of students and faculty are important platforms for learning and creating knowledge together. IASC and IASSA are products of the end of the Cold War when Soviet and Western Arctic scientists could collaborate freely. Research institutions and individual researchers from China and the other Asian nations can and do participate very actively in IASC, IASSA and UArctic. These organizations are also valuable for maintaining Russian-Western research cooperation in the Arctic during political conflict.

There are a couple of Arctic conferences, which because of their size and history also play important roles as platforms of knowledge co-creation and forming epistemic communities. The Arctic Frontiers conference every late January since 2007 in Tromsø has been the long-running large Arctic conference. Arctic Frontiers is a particularly important meeting point between Russia and the West, which reflects the strong competences and networks concerning Russia and the former USSR in Akvaplan-Niva and the University of Tromsø-The Arctic University of Norway. During the Nobel Peace Prize crisis between China and Norway from 2010–2017, the Arctic Frontiers conference was one of the few avenues in Norway where Chinese Arctic researchers would participate.

The other large Arctic conference is the Arctic Circle Assembly every autumn in Reykjavik since 2013. The Arctic Circle Assembly has a very strong North American and global network which represents the international skills and networks of then President Ólafur Ragnar Grímsson (Bertelsen 2015). As on occasions during the Cold War, Iceland has seen its opportunity of being a meeting ground between super powers. The USA, China, Russia and large European countries have been very present at the Arctic Circle Assembly communicating their Arctic policies and strategies.

Sino-Nordic Arctic research ties are particularly well structured and vibrant within the framework of the China-Nordic Arctic Research Center, CNARC. This is a virtual center established in 2013 and based at the Polar Research Institute of China in Shanghai originating in a joint initiative from RANNÍS-The Icelandic Center for Research (Research Council) and PRIC who were joined by the Nordic Institute of Asian Studies at the University of Copenhagen, Fridtjof Nansen Institute (Norway), Norwegian Polar Institute, Swedish Polar Secretariat and Arctic Center at the University of Lapland (Finland), Shanghai Jiao Tong University, Tongji University, Ocean University of China and Shanghai Institutes for International Studies. Recently, Aarhus University, University of Tromsø-the Arctic University of Norway and Dalian Maritime University have joined. CNARC organizes the China-Nordic Arctic Cooperation Symposium (research) with a transdisciplinary business roundtable every second year in China and every second year in the Nordic countries since 2013. Based on personal participant observation, CNARC has traits of an epistemic community where Chinese and Nordic Arctic researchers co-create knowledge, build mutual understanding and increasingly share perceptions of challenges and solutions.

Finally, there are bilateral or trilateral dialogues between the USA, China, Japan, South Korea and Russia, which will be briefly sketched. Shanghai Polar researchers in natural and social sciences from the same institutions as in CNARC have together with American think tank and Polar officials organized two Sino-US Arctic Social Science Forums (sometimes mentioned as China-US Arctic Social Science Forum). The first was in Shanghai 16–17 May 2015, organized by Tongji University and the Center for Strategic and International Studies in Washington DC. The second was at CSIS in Washington DC on the 16–18th of May 2016. There are no public websites for either forum. Presentations and group photos from the last forum reveal the key Chinese and American international politics and policy academics, think tank individuals and officials presenting on their general Arctic international politics research. This research usually from the Chinese side suggests commonality of Chinese and American interests in the Arctic and a legitimate stakeholder role for China in Arctic governance.

Russia and China have a long common border in the Far East and the two countries have a very important relationship for both sides. This relationship is also reflected in mutual research and research cooperation. One example was the Valdai Club conference at East China Normal University in Shanghai in March 2016, which brought together key scholars from both sides. In April 2016, the Arctic ambassadors of China, Japan and South Korea met in Seoul to discuss common Arctic research. In light of the complex relations between these three countries, it is noteworthy if they coordinate their Arctic views and activities. In March 2015, the University of Alaska Fairbanks hosted a Japan-US Arctic Strategy and Policy Workshop. Such a workshop is completely uncontroversial in light of the very close US-Japan alliance and the longstanding connection of University of Alaska Fairbanks with Japan.

Conclusion: Arctic Marine Resource Governance Under Power Transition

The Arctic continues to be an integrated part of the international political, economic and security system, which today is marked by power transition with the return of China to its historical relative weight in the world economy challenging the hegemonic state of the USA. This power transition is the central international policy and research question. This question is also the backdrop for Arctic affairs today, including Arctic marine resource governance, which is particularly clear in the case of the CAO. The Arctic is also potentially a theatre of horizontal escalation of conflicts between the West and post-Soviet Russia. The social and political psychological perspective in international relations emphasizes the importance of information, perceptions, beliefs, judgment and decision for managing such dangerous international processes, whereas a rational actor perspective will assume sufficient information and perception of change on both sides. Transnational epistemic communities of experts co-creating knowledge and sharing beliefs on challenges and solutions can be an important part of managing highly complex and risky processes between states. Science diplomacy can contribute to building such epistemic communities.

There are well-developed and long-standing multilateral Arctic settings for transnational knowledge relations and possible epistemic communities, the Arctic Council working group structure, IASC, IASSA and UArctic. These multilateral settings are key for bringing researchers from the Arctic status quo and the rising Asian countries together – and to provide a Russia-West meeting ground resilient to political crisis. The Arctic Frontiers and Arctic Circle Assembly conferences play similar roles. There is an especially well-developed Sino-Nordic Arctic epistemic community around CNARC and its member institutions.

Bilateral dialogues are less developed, but that is not to be regretted as multilateral settings are preferable for co-creating knowledge and building trust. Sino-American Arctic transnational knowledge relations are not – publicly – well-organized, but the relevant Chinese and American academic, public, private and civil society participants are very familiar with each other from other multilateral academic and policy settings. There have been Sino-US Arctic social sciences fora in Shanghai and Washington DC in 2015 and 2016, where these participants have met. They have presented research, which reflects the two sides sharing their arguments for collaboration and their own legitimate interests.

Transnational knowledge networks contribute to the co-creation of knowledge and mutual understanding and trust that may lead to an epistemic community with shared beliefs on challenges and solutions. Such a development is both a challenge and of value in light of the very complex and potentially dangerous relationship between the USA as existing hegemon and status quo power and China as the rising power – also in the Arctic.

References

Bertelsen, R. G. (2014). The interaction of natural and social systems: how International Relations theory can inform research on Arctic marine invasive species. In L. Fernandez, B. A. Kaiser, & N. Vestergaard (Eds.), *Marine invasive species in the Arctic* (pp. 147–161). København: Nordic Council of Ministers.

Bertelsen, R. G. (2015). The 2nd Arctic Circle assembly: Arctic science diplomacy at work. *The Polar Journal, 5*(1), 240–243.

Bertelsen, R. G., Li, X., & Gregersen, M. H. (2016). Chinese Arctic science diplomacy: An instrument for achieving the Chinese dream? In S. Iglesias Sanchez & E. Conde Perez (Eds.), *Global challenges in the Arctic region: Sovereignty, environment and geopolitical balance* (pp. 442–460). Abingdon: Ashgate.

Borgerson, S. G. (2008). Arctic Meltdown: The Economic and Security Implications of Global Warming. *Foreign Affairs, 87*(2), 63–77.

Callahan, B. (2015). 2015–07-08-last update, *Mearsheimer vs. Nye on the Rise of China* [Homepage of The Diplomat] [Online]. Available: http://thediplomat.com/2015/07/mearsheimer-vs-nye-on-the-rise-of-china/ [2017, 03-21].

Gjørv, G. H., Bazely, D. R., Goloviznina, M., & Tanentzap, A. J. (Eds.). (2014). *Environmental and human security in the Arctic*. London/New York: Earthscan.

Haas, P. M. (1992). Introduction: Epistemic communities and international policy coordination. *International Organization, 46*(1), 1–35.

Hansson, S. (1998). Malm, räls och elektrisitet – skapandet av ett teknologiskt megasystem i Norrbotten 1880–1920 [Iron ore, rail and electricity – creating a technological megasystem in Norrbotten 1880–1920]. In P. Blomkvist & A. Kaijser (Eds.), *Den konstruerade världen – Tekniska system i historiskt perspektiv* (pp. 45–76). Stockholm: Symposion.

Harvey, D. (1989). *The condition of postmodernity: an enquiry into the origins of cultural change*. Oxford/Cambridge, MA: Blackwell.

Heininen, L., & Southcott, C. (Eds.). (2010). *Globalization and the circumpolar North*. Fairbanks: University of Alaska Press.

Herz, J. H. (1950). Idealist internationalism and the security dilemma. *World Politics, 2*(2), 157–180.

Jervis, R. (1997). *System effects: Complexity in political and social life*. Princeton: Princeton University Press.

Lau, R. R. (2003). Models of decision-making. In D. O. Sears, L. Huddy, & R. Jervis (Eds.), *Oxford handbook of political psychology* (pp. 19–59). Oxford: Oxford University Press.

Li, X., & Bertelsen, R. G. 2013. The drivers of Chinese Arctic interests: Political stability and energy and transportation security. In L. Heininen, H. Exner-Pirot, & J. Plouffe (eds.), *Arctic Yearbook 2013* (pp. 1–16). Northern Research Forum; University of the Arctic Thematic Network on Geopolitics and Security. http://www.arcticyearbook.com

Mearsheimer, J. J. (2014). Why the Ukraine crisis is the West's fault: The liberal delusions that provoked Putin. *Foreign Affairs, 93*(5), 77–89.

Nye, J. S. J. (2006). The challenge of China. In S. Van Evera (Ed.), *How to make America safe: New policies for national security* (pp. 73–77). Cambridge, MA: Tobin Project's National Security Working Group.

Nye, J. S., Jr. (2011). *The future of power*. New York: PublicAffairs.

Organski, A. F. K. 1968 [1958]. *World politics* (2nd ed. [rev.] edn.). New York: Knopf.

Pan, M., & Huntington, H. P. (2016). A precautionary approach to fisheries in the Central Arctic Ocean: Policy, science, and China. *Marine Policy, 63*, 153–157.

Sears, D. O. (1975). Political socialization. In F. I. Greenstein & N. W. Polsby (Eds.), *Handbook of political science* (pp. 93–153). Reading: Addison-Wesley Pub..

The Royal Society & AAAS. (2010). *New frontiers in science diplomacy: Navigating the changing balance of power*. London/Washington, DC: The Royal Society/AAAS.

Waltz, K. N. (1979). *Theory of international politics* (1st ed.). New York: McGraw-Hill.

Reshaping Energy Governance in the Arctic? Assessing the Implications of LNG for European Shipping Companies

Ryan Holmes, Darren McCauley, and Nick Hanley

Abstract Future estimates indicate that the reduction of the Arctic ice cap will open up new areas and increase the viability of the region to be increasingly used for international shipping (Liu and Kronbak, J Trans Geo 18(3):434–444. doi:https://doi.org/10.1016/j.jtrangeo.2009.08.004, 2010). The Arctic sea routes and related coastal area are therefore gaining increasing levels of interest, as they become a more attractive alternative for maritime transport. This demand for new infrastructure and development in areas where there has previously been little or none, presents a unique situation to analyze. The increased interest and demand for new development along Arctic sea routes through an environmentally sensitive region make the Arctic an ideal area of which to study the transition toward liquefied natural gas becoming the prominent marine fuel.

We must develop a better understanding of how and under what conditions such a transition will take place and who will make decisions that will influence any such transition. Exploring past and current aspects of maritime and energy governance is an important step in developing an understanding of how a transition towards liquefied natural gas could re-shape our understanding of Arctic governance.

Keywords Arctic • Governance • Energy • LNG • Shipping • Maritime

Introduction

The focus for this chapter is the ongoing transition away from heavy oil based fuels used for maritime transportation towards the utilization of liquefied natural gas as a replacement. Current and forthcoming International Maritime Organization environmental regulations include fuel sulphur content limits, nitrogen oxide emissions limits, and the creation of Emissions Control Areas (ECA's) in

R. Holmes (✉) • D. McCauley • N. Hanley
Department of Geography and Sustainable Development, University of St Andrews,
St Andrews, UK
e-mail: rth3@st-andrews.ac.uk

© Springer International Publishing AG 2018   19
N. Vestergaard et al. (eds.), *Arctic Marine Resource Governance and Development*,
Springer Polar Sciences, https://doi.org/10.1007/978-3-319-67365-3_2

which stricter emissions regulations will be implemented. These are prompting a transition away from the highly polluting heavy oil based fuels traditionally used for maritime transport towards a cleaner burning and more environmentally responsible alternative. Among the practical available alternatives, liquefied natural gas (LNG) is emerging as a preferred option due to several characteristics that give the fuel certain advantages in comparison to that of other alternatives (IMO 2015).

The transition toward liquefied natural gas becoming the prominent fuel used for maritime shipping is likely to have a number of socio-economic costs, as well as benefits. These socio-economic costs and benefits will be manifest in the form of environmental, social, political and economic impacts that will be felt on a wide selection of stakeholders. From an environmental standpoint, the transition to LNG has the potential to result in some positive environmental outcomes, such as diminished pollution levels, decreased risks of environmental disasters and mitigated climate changes consequences. However, the transition to LNG also poses the risk of resulting in some negative environmental outcomes. For instance, if the transition to LNG leads to a larger number of vessels complying with heightened regulations, thus resulting in shipping in protected areas becoming more viable, and traffic and related industrial activity substantially increase in these areas, LNG utilization could have a detrimental environmental effect as a consequence.

From an economic standpoint, the transition to LNG has the potential to spur economic activity and development in certain areas, while reducing it in others. The maritime industry will be confronted with substantial economic costs due to the transition as investments in technology and infrastructure will be required to conform and adapt, particularly in the early stages of the transition. This chapter sets out the key governance challenges involved in managing the transition towards the use of LNG in shipping. For the purposes of this chapter, it focuses on the Arctic region as an invigorating context in which to appreciate the scale of the task confronted by the maritime industry.

Within this context, we will reflect on the governance implications of this transition in shipping from heavy oil to gas as a fuel. We begin with a brief review of previous fuel oil transitions, governance mechanisms, and particularly Arctic governance. We identify three specific governance challenges that emerge from this new context and shipping. The first involves the internal governance of the shipping companies themselves. We reflect on how they can encourage the development of gas based fuel. The second governance challenge is the implication that energy exploitation in the region may increase due to the ease with which shipping companies will be able to deliver hydrocarbons from the region, both in terms of melting ice and the less polluting future of LNG. Current restrictions on shipping in the area may transform very quickly. The third governance challenge is how we may in the future include shipping companies as a stakeholder in the region. Currently, we treat companies in shipping as legal entities to be regulated.

Energy Transitions in Shipping Fuel: From Oil to . . .

At the dawning of the twentieth century, the potential of petroleum powered maritime transportation began to attract growing attention. A century after inventor Rudolf Diesel patented his revolutionary compression ignition engine in 1892, his design would be the predominant engine type powering the world's merchant vessels (Griffiths 1995). Petroleum offered several advantages over coal as a fuel for maritime transportation. Engineers and innovators began experimenting with the potential of petroleum as a maritime transportation fuel around the turn of the twentieth century. Significant advancements began to take place in the quest for naval improvements in the years prior to World War I. Naval strategists of the era began to view the utilization of petroleum as a fuel that could offer naval fleets advantages such as improvements in speed, a more practical refueling procedure, and a considerable reduction in weight that would thus allow for the installation of increased protective armor as well as additional armaments (Hugill 2014). The transition to oil assisted the United States in being able to enhance its naval power in the Pacific. The transition to oil allowed for the utilization of oil reserves located in California as opposed to coal. Previously, coal was required to be shipped from supply sources located in distant regions due to the type of coal found in the western United States being regarded as unsuitable for use in steamship engines (Painter 2012). Wartime negotiations and policies such as the Lend-Lease agreement between the United States and Great Britain are examples of policies that were linked to the increased importance of petroleum as a vital source of energy (Venn 2012).

We are now in an era of uncertainty with regards to the substitution of oil, and what role governance should play in this substitution. Nuclear provides an interesting example. Beginning in the very early stages of nuclear technological development, the prospect of using nuclear energy for the purpose of maritime transportation was examined. For example, the United States Navy directed much of the early research related to nuclear technology (Hultman 2011). Pressurized water reactors were initially designed in connection with the Unites States Navy, for the purpose of powering submarine propulsion in an effort to increase the amount of time it was possible for a submarine to remain submerged without refueling, and in 1955 the world's first nuclear powered submarine, the USS *Nautilus* was launched (Oka et al. 2014).

Following the pioneering developments of nuclear powered submarine applications, the use of nuclear powered maritime propulsion expanded to also be utilized by surface vessels and has since been adopted in such a capacity not only by major naval fleets throughout the world, but also for such applications as icebreaker vessels, the first of which was put into service by the USSR in 1959 (Hirdaris et al. 2014). Russian nuclear icebreakers operating along the Northern Sea Route serve a crucial role in the economic development of the Arctic, and nuclear icebreakers have the important advantage over diesel petroleum powered icebreakers by being able to operate over significantly longer periods of time without the need for refueling (Bukharin 2006).

Historic energy transitions provide many lessons that can serve as useful guidelines for policy makers facing the challenges of future energy transitions such as the transition towards LNG. Lessons from the transition from organic sources of energy to coal demonstrate the importance of sound environmental policy and regulation with effective enforcement and compliance mechanisms that are administered in a fair and just manner. International trade is a fundamental aspect of the transition towards LNG as a maritime propulsion fuel, particularly within the Arctic region, where the prospect of potential increases in trade and related development activity taking place within the region present the potential for significant impacts not only for the Arctic region but also well beyond.

Governance

Governance can be described as a process that moves beyond the conventional model of state controlled organizational units implementing collectively binding decisions through the use of established bureaucracy towards one which also encompasses the influence of private third sector actors within society that participate in the policy process through actions that steer policy decisions (Treib et al. 2007). Therefore, governance can be viewed as including not only the traditional state, but also incorporating other parties such as businesses and nongovernmental organizations into the political decision making process through the use of steering and influence (Jordan 2008). Farazmand (2013) points to the concept of governance as essentially expanding beyond the established state to also include societal organizations and the private sector, and therefore empowering these parties to promote their interest and negotiate their differences along with those of the state.

Although the two are certainly closely associated, governance should not be confused with government, as the two terms are different concepts and certainly not simply different terms for the same theory. Florini and Sovacool (2009) attempt to clarify this difference by pointing out that governments can be considered as merely one of many aspects of the overall governance concept in which rules are set and enforced by groups of people in an effort to realize desired conclusions. Hence, while traditional formal governments are a predominantly important and significant part of the governance process, they also share in this process to varying degrees with non-state participants from differing segments of society. As Jordan (2008) indicates, government tends to relate to actions that are based in recognized formal authority, while governance on the other hand relates to actions based on shared goals that also include nongovernmental and informal institutions. Young (2012) depicts governance as efforts intending to steer sociological systems towards desirable directions through the establishment of rules and decision-making processes.

Another aspect of governance is that it can transcend specifically defined geographic areas across the local, national and international spectrum (Jordan 2008). This aspect of governance is particularly important in the recent era of increased globalization. Farazmand (2013) lists factors such as technological innovation, the fall of the former Soviet Union, and the expanding role of the United Nations' agencies as reasons that globalization has intensified in recent history, and along with other factors, has led to an increased transformation of governance beyond the traditional state. This is a transformation characterized by the increasing influence of non-state parties.

Biermann (2010) notes a shift from intergovernmental organizations towards increased public-private cooperation taking place over the past decades, which has been characterized by an increasing number on non-state actors becoming formally involved in norm setting and implementation throughout the world. Driessen et al. (2012) states that, despite recent shifts in modes of governance, modes of governance tend to build upon one another instead of completely replacing one another, and further, traditional hierarchal structures are often found alongside new modes of governance.

Arctic and Maritime Governance

A particularly intriguing example of maritime governance in this transition to LNG or alternative fuels is the current issue of development taking place in the Arctic. Increasing pressures from forces such as climate change and globalization are driving a transformation in the Arctic and are attracting an increasing amount of interest in the Arctic from parties such as shipping, oil and gas exploration, commercial fishing and tourism that are likely to not only transform but also challenge the governance of the Arctic region (Young 2012).

As a consequence of climate change, the ice and snow cover of the Arctic has been significantly subsiding, and as a result the region has become increasingly more accessible and of interest to a collection of different activities (Smits et al. 2014). As Young (2012) points out, while the Arctic once aroused little international interest on the world stage, in light of recent development, it now is attracting a rapidly increasing amount of both political and economic attention from the international community. The emergence of this increased interest and associated activities poses not only numerous economic and geopolitical opportunities for the Arctic region, but also the potential for dire environmental and social consequences if the risk from this mounting interest in the development of the Arctic region is not carefully managed going forward.

The unique distinctive physical and political characteristics of the Arctic region have led to it having a rather intriguing arrangement of governance. A principal player in Arctic governance is the Arctic Council. The Arctic Council was established in 1996 by the Ottawa Declaration to manage sustainable development and environmental protection of the Arctic through coordination and cooperation among Arctic states as well as the indigenous peoples living in the Arctic region (Smits et al. 2014). The membership of the Arctic Council consists of the eight states with territory located within the Arctic: the United States, Canada, Finland, Iceland, Norway, Russia, Sweden, and Denmark (Greenland), as well associations representing the indigenous residents of the Arctic region (Stokke 2013).

An interesting feature of the Arctic Council is the ability of non-Arctic state stakeholders, such as other non-Arctic states and Nongovernmental Organizations, to apply for observer status with the Council (Smith et al. 2014). Some examples of organizations that have applied for and been granted observer status by the Arctic Council include the non-Arctic states of China, the Netherlands, and Japan, as well as a number of nongovernmental organizations such as the World Wildlife Fund. Other organizations such as the European Union have been slow to gain observer status, despite having applied for it and desiring to obtain it (Smits et al. 2014). The increasing collection of organizations eager to obtain observer status with the Arctic Council, particularly powerful non-Arctic states and international organizations, is an indication of the status that the Arctic region has gained in political circles, and also demonstrates the important role the Arctic Council plays in the governance of the region (Smits et al. 2014).

Nongovernmental organizations have the potential to play an extraordinary part in the governance of the Arctic. Although environmental nongovernmental organizations such as Greenpeace have struggled with low levels of support in the Arctic, particularly in Greenland, due to past endeavors such as anti-whaling efforts, they now have the ability to organize the population of the region and to also act as monitors of the regions oil and gas resource development, and therefore increase their influence and play an critical role in the governance of the Arctic region (Smits et al. 2014).

Another aspect that makes the governance of the Arctic region unique and interesting is the relationship between Denmark and Greenland. For instance, while Denmark officially holds a membership seat on the Arctic Council, it is Greenland, who established Self-Government in 2009, that is actually the most active participant of the two when it comes to the Arctic Council, with Denmark usually supporting and following Greenland's position on most issues (Smits et al. 2014). However, the increased utilization of their natural resources, particularly those of the oil and gas sector, is seen by many as a means for Greenland to become financially independent and will certainly play a factor in the governance of Greenland and the Arctic region in the future (Smits et al. 2014).

The Energy Imperative: Arctic Oil and Gas

Harsem et al. (2011) point to three major factors that will influence the expansion of oil and gas development in the Arctic region as being climate change, economic and market conditions, and the level of government encouragement by Arctic states. Of all these factors, climate change is perhaps the one most associated with recent issues in the Arctic. Studies of climate models have indicated that global warming will be even more enhanced in high northern latitudes. It is also predicted that the Arctic will be the location of the most dramatic and rapid changes occurring over the next century (Ho 2010).

While global warming and melting ice might facilitate the development of the Arctic's oil and gas resources by making these resources easier to reach and exploit, climate change also presents a series of challenges to the development of Arctic oil and gas. An increase in the frequency and severity of extreme weather conditions resulting from climate change, such as hurricanes, would have dire effects on oil and gas developments in the Arctic and present the possibility of devastating events ranging from costly production and transportation disruptions to disasters such as oil spills (Harsem et al. 2011). These potential risks will certainly factor into the rate at which oil and gas development in the Arctic progresses.

While climate change and melting ice are often presented as the main factors influencing increased interest in Arctic oil and gas developments, Bennett (2014) draws attention to arguments contending that it is rather energy prices and a desire to secure resources that have been the actual facilitators of the recent heightening of interest in Arctic oil and gas development. The Fukushima nuclear disaster leading to an increase in oil and gas purchases by Japan is an example presented by Bennett (2014) as one illustration of a recent event that has led to an increased interest in oil and gas development in the Arctic. Harsem et al. (2011) assert that in the future, global economic conditions will be the most important determinant of oil and gas developments in the Arctic, and point towards the worldwide effects that the financial crisis of 2008 had on the price and demand of energy as well as energy related investments as evidence in support of this position.

Arctic Shipping

An aspect of the Arctic that is attracting increased international attention is the potential for increased utilization of Arctic shipping lanes. Future estimates indicate that the reduction of Arctic ice cap will open up new areas and increase the viability of the region to be increasingly used for international shipping (Liu and Kronbak 2010). According to Sakhuja (2014), the two most practical Arctic shipping routes are the Northern Sea Route and the Northwest Passage. Via the Arctic, large bulk carriers can significantly reduce the distance between Asia, Europe, and North America by navigating the Northern Sea Route or the Northwest Passage and the increased melting of Arctic sea ice poses the potential for an expanded navigation season along the routes (Hong 2012).

Running between the Atlantic and Pacific along the Russian coast, the Northern Sea Route ranges between 2100 and 2900 nautical miles depending on the distribution of sea ice, and the Northern Sea Route is part of the shortest connection between Northeast Asia and Northern Europe (Liu and Kronbak 2010). Examples presented by Hong (2012) of the potential reductions in sailing distances afforded by the Arctic routes include the sailing distance of a voyage between Rotterdam and Yokohama via the Northern Sea Route instead of the Suez Canal being reduced by 40% and the sailing distance of a voyage between Rotterdam and Seattle via the Northwest Passage instead of the Panama Canal being reduced by 25%.

The prospect of increased shipping activity along Arctic routes also presents a collection of concerns and considerations that must be addressed. Ho (2010) lists increased infrastructure investments and the establishment of expanded marine services focused on safety and environmental responsibility throughout the region, as steps that are necessary before the Arctic sea routes can be reliably used on a large scale. Liu and Kronbak (2010) discuss various construction and equipment standards such as hull thickness and structural support requirements that are necessary for ships to be qualified as an ice class vessel. Certainly these issues will be taken into consideration among others factors by the maritime community and determine how quickly the utilization of Arctic sea routes increases in the future.

As the interest in Arctic development activities increases, particularly the prospect of substantially increased shipping activity taking place in the Arctic, the need for specialized compulsory shipping regulations that address the unique challenges and concerns related to the Arctic becomes ever more essential. In recognition of the complicated challenges the Arctic region faces due to the increased interest in Arctic shipping, the International Maritime Organization has initiated the development of a mandatory international code of safety for ships that operate in Arctic waters, which would compliment guidelines and regulations that are already in place (Jabour 2014). The Polar Code addresses such issues as vessel design, construction and equipment, search and rescue procedures, training and environmental protection and will focus on specific risks associated with operating in Arctic waters (Hartsig et al. 2012). The establishment of this code plays a crucial role in the future development of the Arctic region. The work and collaboration between the Arctic Council and the IMO in the development of the Polar Code, which was entered into force on 1 January 2017, is an excellent example of how governance institutions are working together on important issues (Arctic Council 2016; IMO 2017).

The discharge of air pollution resulting from international shipping has serious negative effects that are harmful to both health and the environment. As the impact of these detrimental effects become more clearly understood, efforts aimed at the abatement of these emissions have received increased attention, particularly in the form of recent regulatory actions. Until recent regulatory efforts began inciting the need for a viable alternative, most large vessels engaged in international shipping burned heavy fuel oil which is a residual by-product of the refining process (Bloor et al. 2014).

The emissions generated from the use of heavy fuel oil by the shipping sector discharge into the air large volumes of SO_x and NO_x, which have been shown to be extremely harmful to crops, forests and the ocean as a result of acidification, as well as fine particulate matter, which has been shown to be a cause of serious health issues such as lung disease and coronary illness (Bloor et al. 2014). Despite maritime transport having a favorable emissions performance in comparison to that of land and air transport, ocean going shipping is still responsible for an estimated 15% of anthropogenic NO_X emissions and 5–8% of SO_x emissions globally (Viana et al. 2014). The emission of these harmful substances by ships therefore has a serious impact on world health; with one study by Corbett et al. (2007) estimating that emissions from shipping contributed to 64,000 premature deaths worldwide in 2002 (Bloor et al. 2014). Black Carbon (BC) emissions generated from shipping are of a particular concern for the Arctic region, as these pollutants have been shown to reduce surface albedo as a result of absorbing light (Aliabadi et al. 2015).

New Governance Stakeholders: Shipping Companies

Of the numerous stakeholders involved in the transition to LNG becoming the prominent marine fuel, among the most involved and impacted are vessel owning and operating companies. Within the focus area of the Arctic region, there is a diverse collection of shipping companies, which vary in characteristics ranging from their purpose, the type of vessels employed, to certain specialties and niche services provided. The extreme physical characteristics of the Arctic lead to shipping activities within the region being rather unique in comparison to maritime activity elsewhere in the world. An overview of a collection of some prominent types of shipping companies and their respective endeavors in the Arctic region is an important step in gaining an understanding of the ramifications of the transition towards LNG for future governance challenges in the Arctic. Although the small collection of companies discussed below represents merely a fraction of the vessels operating in the Arctic, with the total number of vessels active in Arctic waters estimated at 6000 in 2004, they represent an interesting insight to a category of principal actors in Arctic governance (Arctic Council 2009).

Of the shipping firms that are active in the Arctic, some companies' principal areas of operation are primarily within the Arctic and the surrounding far northern regions. One such company is Nunavut Eastern Arctic Shipping Inc. It provides cargo and supply services to the Arctic with a fleet of multi-purpose container vessels and supporting equipment (NEAS 2015). It is headquartered in Iqaluit, Canada, and is a majority Inuit owned company. The Canadian shipping company Fednav is headquartered in Montreal, with European offices based in London. Fednav has a significant presence in Arctic maritime operations. It has the world's largest fleet of ice-class bulk carriers with which they connect the Great Lakes and Canadian Arctic with markets throughout the world (Fednav 2015). It is a large international company with a history of expertise concerning Arctic maritime operations.

An example of companies operating specialty vessels designed for a specific purpose in Arctic waters are seismic data acquisition vessels such as those operated by London based WesternGeco, which is a segment of the global oil field services company Schlumberger. According to the company's website, their new Amazon-class seismic data acquisition vessel was designed to Polar-Class 7 specifications with the aim of enhanced capabilities of conducting seismic research operations in the Arctic region (WesternGeco 2015).

Eimskipafélag Íslands is headquartered in Reykjavík, Iceland. Eimskipafélag Íslands, commonly referred to as Eimskip, is Iceland's oldest shipping company and has offices in eighteen locations throughout the world, including the United Kingdom (Eimskip 2015). Greenlandic company Royal Arctic Line is headquartered in the capital city of Nuuk. According to the company's website, Royal Arctic Line is owned by the government of Greenland and provides shipping services to and from Greenland with a fleet of 10 vessels specially designed for Arctic operations (Royal Arctic Line 2015). It is a wholly government owned firm.

Implications for Energy Governance

The increased role of shipping companies in the present and future governance of the Arctic demands a new perspective on how institutions, policies and governance solutions are shaped. Up to this point, policy makers have focused on the developers and those affected by their work. Normally, a governance assessment of the Arctic would include a focus on unfair development in relatively untouched parts of vulnerable regions (McCauley et al. 2016; Sidortsov and Sovacool 2015). At an international level, we would focus on the Arctic Council. At a national level, our assessment would include national policies towards including various stakeholders in agreeing on a plan of action. Local initiatives would emerge from communities who are involved at the hard face of the decisions made internationally and nationally. But what do we do when the key stakeholder in focus is not fixed in one point? Shipping companies are indeed this form of stakeholder.

The Polar Code and other international regulations have been the vehicles through which we have dealt with shipping companies with regards to the Arctic (Chang et al. 2015). The drop in the price of natural gas has now made shipping companies a more present actor in the region. Of course, oil and gas prices rise and fall over time. The new shale gas revolution does however threaten a more long-term institutionalised low-level cost for gas. The first governance question for shipping companies is not the role that they should play in international, national, or local governance. It is rather how they govern the transition away from heavy oil based fuel towards LNG. Shipping companies must acknowledge that the long-term future for global shipping routes is indeed gas, not oil reliant. We argue that the Arctic area of the world offers one of the first insights into how the shipping companies might manage this transition away from oil towards gas.

The second governance implication of this shift from gas to oil for shipping companies is indeed the impact upon present and future expectations of hydrocarbon exploitation in the region. It has been common to suggest that hydrocarbons are no longer easily accessible within the region (Kennicutt et al. 2014). We need to also appreciate that the delivery of resources for exploiting these opportunities are central to understanding future energy development in the region. Moreover, the resources that are exploited need them to be delivered to the communities that will use them (Krivitsky et al. 2001). At the very heart of this governance challenge sits the shipping companies. The present barrier from their perspective is the anonymous damage generated by heavy oil fuel use. If the future does not include oil and terms of transport offers more opportunities for shipping companies to be involved in future energy exploitation.

Shipping in the region cannot be reduced to simply a passage from Asia to the West (Aporta 2009). It is evident that it will equally include a greater role for shipping in helping to exploit and deliver hydrocarbons. With this in mind, the third governance implication is the necessity to include shipping companies as separate stakeholders in the region. At an international level, shipping is understood as a practice that needs to be regulated. This is a narrow view of a shipping company's role. We need to find an institutionalised space for shipping companies to be involved in policy decisions in the region. The Arctic Council could include such stakeholders. They could encourage the involvement of indigenous companies that are involved in transportation for example. This would provide extremely important logistical advice for the development of infrastructure. We remain too reliant on the three stakeholder groups of energy companies, nation-states, and local communities. The Arctic Economic Council is perhaps a more relevant venue.

Solutions at a national or local level are a little more difficult. The very nature of international shipping means that national or local governance is constrained by a lack of affiliation with communities (Aporta 2009). Many companies will not be from any of the Arctic nation-states. Shipping lanes will pass through communities, but without opportunity for people to input into whether it should be allowed or can offer additional benefits (Zolotukhin and Gavrilov 2011). We do not foresee the emergence of innovative forms of local or national governments that will be able to successfully circumnavigate this challenge. This places more emphasis on the need to think creatively at the international level. A new shipping council could emerge depending upon the future patterns of activities in the region. For now, the primary governance challenge for shipping companies is to successfully transition from oil to gas. As this is currently taking place, we need to begin to think about implications for governance in the Arctic. We hope that we have offered some initial reflection on this potential future.

Conclusion

An analysis of previous energy transitions points to the ability of energy transitions to act as an instrument of economic development and growth. Future energy transitions offer the possibility, just as past transitions have, to remove barriers and limits to economic growth. By offering the potential to free energy users form existing constraints related to energy use and dependence on particular sources, future energy transitions can provide new alternatives and prospects to society that can have meaningful global impacts for the future, not only in economic terms but also in political and social outcomes as well.

The nature of the Arctic region presents a unique set of challenges that stakeholders must face when engaging in development within the region. How well the different stakeholders cooperate and coordinate their efforts to address these challenges will play a large role in determining the success of efforts to increase development in the Arctic and prepare for energy transitions in shipping and beyond. Kao et al. (2012) point to the 2011 signing of the Agreement on Cooperation on Aeronautical and Maritime Search and Rescue in the Arctic as representing an example of Arctic states reacting to these challenges. A lack of infrastructure, extreme weather conditions, ice and remoteness make petroleum development activities exceptionally challenging and thus the risk of oil spills is of particular concern in the Arctic region (Knol and Arbo 2014). The bilateral oil spill response established by Norway and Russia is an example of an effective system of Arctic states cooperating to address concerns of risks of petroleum development related risks in the region (Sydnes and Sydnes 2013).

The development of the Arctic region itself presents, therefore, an opportunity to encourage an energy transition in shipping fuel. Its unique environment, both physical and political, can lead to environmental, economic and social imperatives, which may drive the adoption of LNG. Future research should quantitatively investigate both the costs and benefits of such a transition.

References

Aliabadi, A., Staebler, R., & Sharma, S. (2015). Air quality monitoring in communities of the Canadian Arctic during the high shipping season with a focus on local and marine pollution. *Atmospheric Chemistry and Physics, 15*(5), 2651–2673.

Aporta, C. (2009). The trail as home: Inuit and their pan-arctic network of routes. *Human Ecology, 37*(2), 131–146.

Arctic Council. (2009). *Arctic marine shipping assessment 2009 report*. Retrieved from http://www.pame.is/index.php/projects/arctic-marine-shipping/amsa

Arctic Council. (2016). *The Arctic council: A forum for peace and cooperation*. Retrieved from http://www.arctic-council.org/index.php/en/our-work2/8-news-and-events/415-20th-anniversary-statement

Bennett, M. M. (2014). North by northeast: Toward an Asian-Arctic region. *Eurasian Geography and Economics, 55*(1), 71–93.

Biermann, F. (2010). Beyond the intergovernmental regime: Recent trends in global carbon governance. *Current Opinion in Environmental Sustainability, 2*(4), 284–288. https://doi.org/10.1016/j.cosust.2010.05.002

Bloor, M., Sampson, H., Baker, S., & Dahlgren, K. (2014). The instrumental use of technical doubts: Technological controversies, investment decisions and air pollution controls in the global shipping industry. *Science and Public Policy, 41*(2), 234–244. https://doi.org/10.1093/scipol/sct050

Bukharin, O. (2006). Russia's nuclear icebreaker fleet. *Science and Global Security, 14*(1), 25–31.

Chang, K., He, S., Chou, C., Kao, S., & Chiou, A. (2015). Route planning and cost analysis for travelling through the arctic northeast passage using public 3D GIS. *International Journal of Geographical Information Science, 29*(8), 1375–1393.

Corbett, J. J., Winebrake, J. J., Green, E. H., Kasibhatla, P., Eyring, V., & Lauer, A. (2007). Mortality from ship emissions: A global assessment. *Environmental Science & Technology, 41*(24), 8512–8518.

Driessen, P. P., Dieperink, C., Laerhoven, F., Runhaar, H. A., & Vermeulen, W. J. (2012). Towards a conceptual framework for the study of shifts in modes of environmental governance–experiences from the netherlands. *Environmental Policy and Governance, 22*(3), 143–160.

Eimskip. (2015). *Eimskip*. Retrieved from http://eimskip.is/EN/Pages/default.aspx

Farazmand, A. (2013). Governance in the age of globalization: Challenges and opportunities for south and southeast asia. *Public Organization Review, 13*(4), 349–363.

Fednav. (2015). *Fednav*. Retrieved from http://www.fednav.com/en

Florini, A., & Sovacool, B. K. (2009). Who governs energy? The challenges facing global energy governance. *Energy Policy, 37*(12), 5239–5248. https://doi.org/10.1016/j.enpol.2009.07.039

Griffiths, D. (1995). British shipping and the diesel engine: The early years. *The Mariner's Mirror, 81*(3), 313–331.

Harsem, Ø., Eide, A., & Heen, K. (2011). Factors influencing future oil and gas prospects in the arctic. *Energy Policy, 39*(12), 8037–8045. https://doi.org/10.1016/j.enpol.2011.09.058

Hartsig, A., Frederickson, I., Yeung, C., & Senner, S. (2012). Arctic bottleneck: Protecting the bering strait region from increased vessel traffic. *Ocean & Coastal LJ, 18*, 35.

Hirdaris, S. E., Cheng, Y. F., Shallcross, P., Bonafoux, J., Carlson, D., Prince, B., & Sarris, G. A. (2014). Considerations on the potential use of nuclear small modular reactor (SMR) technology for merchant marine propulsion. *Ocean Engineering, 79*, 101–130. https://doi.org/10.1016/j.oceaneng.2013.10.015

Ho, J. (2010). The implications of arctic sea ice decline on shipping. *Marine Policy, 34*(3), 713–715. https://doi.org/10.1016/j.marpol.2009.10.009

Hong, N. (2012). The melting arctic and its impact on China's maritime transport. *Research in Transportation Economics, 35*(1), 50–57. https://doi.org/10.1016/j.retrec.2011.11.003

Hugill, P. J. (2014). Petroleum supply, marine transportation technology, and the emerging international order of the post world war one period. In *The global politics of science and technology* (Vol. 1, pp. 141–159). Berlin: Springer.

Hultman, N. E. (2011). The political economy of nuclear energy. *Wiley Interdisciplinary Reviews: Climate Change, 2*(3), 397–411. https://doi.org/10.1002/wcc.113

IMO. (2015). *Air pollution*. Retrieved from http://www.imo.org/en/OurWork/Environment/PollutionPrevention/AirPollution/Pages/Air-Pollution.aspx

IMO. (2017). *Polar code*. Retrieved from http://www.imo.org/en/MediaCentre/HotTopics/polar/Pages/default.aspx

Jabour, J. (2014). Progress towards the mandatory code for polar shipping. *Australian Journal of Maritime & Ocean Affairs, 6*(1), 64–67. https://doi.org/10.1080/18366503.2014.888135

Jordan, A. (2008). The governance of sustainable development: Taking stock and looking forwards. *Environment and Planning C, Government & Policy, 26*(1), 17.

Kao, S., Pearre, N. S., & Firestone, J. (2012). Adoption of the arctic search and rescue agreement: A shift of the arctic regime toward a hard law basis? *Marine Policy, 36*(3), 832–838. https://doi.org/10.1016/j.marpol.2011.12.001

Kennicutt, M. C., Chown, S. L., Cassano, J. J., Liggett, D., Peck, L. S., Massom, R., et al. (2014). A roadmap for Antarctic and Southern Ocean science for the next two decades and beyond. *Antarctic Science, 27*(01), 3–18.

Knol, M., & Arbo, P. (2014). Oil spill response in the arctic: Norwegian experiences and future perspectives. *Marine Policy, 50, Part A*(0), 171–177. https://doi.org/10.1016/j.marpol.2014.06.003

Krivitsky, S. V., Tsvetsinsky, A. S., Mirzoev, D. A., & Krivitsky, A. S. (2001). Oil and gas exploration at the arctic shore: Social issues. *Proceedings of the Eleventh (2001) International Offshore and Polar Engineering Conference, I*, 661–664.

Liu, M., & Kronbak, J. (2010). The potential economic viability of using the northern sea route (NSR) as an alternative route between Asia and Europe. *Journal of Transport Geography, 18*(3), 434–444. https://doi.org/10.1016/j.jtrangeo.2009.08.004

McCauley, D., Heffron, R., Pavlenko, M., Rehner, R., & Holmes, R. (2016). Energy justice in the Arctic: Implications for energy infrastructural development in the Arctic. *Energy Research & Social Science.* https://doi.org/10.1016/j.erss.2016.03.019

NEAS. (2015). *Neas.* Retrieved from http://www.neas.ca/

Oka, Y., Uchikawa, S., & Suzuki, K. (2014). Light water reactor design. In *Nuclear reactor design* (pp. 127–229). Tokyo: Springer.

Painter, D. S. (2012). Oil and the American century. *Journal of American History, 99*(1), 24–39.

Royal Arctic Line. (2015). Royal Arctic Line. Retrieved from http://www.royalarcticline.com/.

Sakhuja, V. (2014). The polar code and arctic navigation. *Strategic Analysis, 38*(6), 803–811. https://doi.org/10.1080/09700161.2014.952943

Sidortsov, R., & Sovacool, B. (2015). Left out in the cold: Energy justice and Arctic energy research. *Journal of Environmental Studies and Sciences, 5*(3), 302–307. https://doi.org/10.1007/s13412-015-0241-0.

Smits, C. C., van Tatenhove, J. P., & van Leeuwen, J. (2014). Authority in Arctic governance: Changing spheres of authority in Greenlandic offshore oil and gas developments. *International Environmental Agreements: Politics, Law and Economics, 14*(4), 329–348.

Stokke, O. S. (2013). Regime interplay in Arctic shipping governance: Explaining regional niche selection. *International Environmental Agreements: Politics, Law and Economics, 13*(1), 65–85.

Sydnes, A. K., & Sydnes, M. (2013). Norwegian–Russian cooperation on oil-spill response in the Barents Sea. *Marine Policy, 39*(0), 257–264. https://doi.org/10.1016/j.marpol.2012.12.001

Treib, O., Bähr, H., & Falkner, G. (2007). Modes of governance: Towards a conceptual clarification. *Journal of European Public Policy, 14*(1), 1–20.

Venn, F. (2012). The wartime 'special relationship'? From oil war to Anglo-American Oil Agreement, 1939–1945. *Journal of Transatlantic Studies, 10*(2), 119–133.

Viana, M., Hammingh, P., Colette, A., Querol, X., Degraeuwe, B., Vlieger, I. d., & van Aardenne, J. (2014). Impact of maritime transport emissions on coastal air quality in Europe. *Atmospheric Environment, 90*, 96–105. https://doi.org/10.1016/j.atmosenv.2014.03.046

WesternGeco. (2015). *Westerngeco.* Retrieved from http://www.slb.com/services/seismic/seismic_acquisition.aspx

Young, O. R. (2012). Arctic tipping points: Governance in turbulent times. *Ambio, 41*(1), 75–84.

Zolotukhin, A., & Gavrilov, V. (2011). Russian arctic petroleum resources. *Oil & Gas Science and Technology–Revue d'IFP Energies Nouvelles, 66*(6), 899–910.

Part II
Fishery

Regulating Fisheries in the Central Arctic Ocean: Much Ado About Nothing?

Rosemary Rayfuse

Abstract In July 2015 the five Arctic Ocean coastal states adopted the Oslo Declaration Concerning the Prevention of Unregulated High Seas Fishing in the Central Arctic Ocean in which they voluntarily agreed to refrain from commercial fishing in the Central Arctic Ocean unless and until appropriate science based management measures are in place. Like the Ilulissat Declaration before it, the Oslo Declaration was both hailed as a major achievement and criticised as an act of Arctic exceptionalism. This chapter interrogates the claim of Arctic exceptionalism and demonstrates that the Oslo Declaration, by itself, constrains neither the rights and interests of the Arctic Ocean coastal states nor the rights and interests of the rest of the international community. Of greater import will be the outcome of the subsequent 'Broader Process' negotiations the Declaration has spawned involving non-Arctic Ocean states. Given the uncertainties and extremely limited scientific knowledge regarding existing and potential Central Arctic Ocean fisheries resources, particularly when combined with the current lack of activity in the area, these negotiations provide a valuable opportunity to implement a truly precautionary approach to their future conservation and management based on sound science and modern international fisheries management principles and practices.

Keywords High seas • High seas fisheries • Central Arctic Ocean • Oslo Declaration • Broader process • Arctic Ocean governance • Arctic fisheries • Fisheries management • Fisheries conservation and management

R. Rayfuse (✉)
UNSW Sydney, NSW, Australia

Lund University, Lund, Sweden
e-mail: r.rayfuse@unsw.edu.au

University of Gothenburg, Gothenburg, Sweden
e-mail: r.rayfuse@unsw.edu.au

© Springer International Publishing AG 2018
N. Vestergaard et al. (eds.), *Arctic Marine Resource Governance and Development*,
Springer Polar Sciences, https://doi.org/10.1007/978-3-319-67365-3_3

Introduction

An object of human desire and endeavour, the Arctic has long fascinated and perplexed. While the history of human habitation in the Arctic stretches back more than twelve thousand years (Vaughan 1994), thanks largely to its perennial ice cover, the waters of the high Arctic Ocean have remained inaccessible to all but a few hardy scientists and adventurers and, in recent years, the nuclear powered submarines of a few states. However, at least since the 1980s, international interest in the Arctic has been growing, fuelled by the prospects of a warming and climate changed Arctic delivering up its potentially vast and as yet untapped resources. Even before Russian scientists planted the now infamous titanium Russian flag on the seabed at the North Pole (Blomfield 2007), other states were eyeing-off the opening of new trans-polar shipping routes between Europe and Asia and salivating over the prospect of a new resource bonanza, predicted by some, at least, to lead to new scrambles for wealth and power, the revival of international security tensions and, possibly, international conflict in the Arctic region (see, e.g., Reynolds 2007; Cressey 2007; Borgerson 2008; Sale and Potapov 2010). Given the near fever-pitch of international interest in the Arctic, it was perhaps hardly surprising that the five Arctic Ocean coastal states, Canada, Denmark in respect of Greenland, Norway, Russia and the United States (the Arctic 5 or A5) felt compelled in 2008 to adopt the Ilulissat Declaration in which they reminded the international community that there already exists an 'extensive international legal framework [that] applies to the Arctic Ocean'. There was thus 'no need to develop a new comprehensive international legal regime to govern the Arctic Ocean'.

The Ilulissat Declaration raised the ire of the other member states of the Arctic Council who felt the A5 were undermining it (Molenaar 2016a, p. 449). Others suggested that, in asserting their special role as 'stewards' of the Arctic, the A5 were asserting rights over the Arctic Ocean they might not possess (Young 2016, p. 274). However, like it or not, the A5 were correct. The international law of the sea, including the 1982 United Nations Convention on the Law of the Sea (Law of the Sea Convention, or LOSC) together its two implementing agreements, the 1994 Implementing Agreement on Part XI of the LOSC (Part XI IA) and the 1995 Fish Stocks Agreement (FSA) does, indeed, provide an extensive legal framework for the governance of the earth's oceans – including the Arctic Ocean. Coastal states such as the A5 have sovereignty over their territorial seas, subject only to the right of other states to innocent passage. Within the exclusive economic zone coastal states have jurisdiction over the conservation and management of living and non-living resources and their jurisdiction extends to their continental shelf, even where that shelf physically extends beyond 200 nautical miles. Any interests that the international community – including the non-Arctic Ocean coastal states in the Arctic Council – might have in the Arctic Ocean are thus limited to the rights of innocent passage in the territorial sea, freedom of navigation in the exclusive economic zone and the rights to navigate, fish, lay submarine cables and pipelines, construct artificial islands and conduct marine scientific research in areas beyond national jurisdiction, these are activities all governed already by a plethora of international legal regimes. Moreover, the Arctic Council does not concern itself with matters of sovereignty. In short, both the calls by the international community for some sort of comprehensive Arctic agreement and the adverse reactions by other states to the Ilulissat Declaration were misconceived.

Nevertheless, while extensive, the law of the sea is not exhaustive – as illustrated by current international negotiations on a possible regime for the protection of marine biodiversity in areas beyond national jurisdiction (for information see, e.g. ENB 2016). In the Arctic context, in particular, one significant gap that has been identified is the emerging need for an effective regional fisheries conservation and management regime, applicable to the high seas area of the Central Arctic Ocean which lies beyond the national jurisdiction of the A5 (Rayfuse 2007; Koivurova and Molenaar 2009). In July 2015 the A5 adopted the Oslo Declaration Concerning the Prevention of Unregulated High Seas Fishing in the Central Arctic Ocean (Oslo Declaration) in which they agreed voluntarily to refrain from commercial fishing in the Central Arctic Ocean unless and until appropriate science based management measures are in place. Like the Ilulissat Declaration before it, the Oslo Declaration once again caused consternation among other states worried that their rights and interests in the Arctic, including in access to and long-term conservation and sustainable management of Arctic fisheries resources, were being ignored. Iceland, in particular, objected to having been excluded from the discussions, apparently on the basis that fish species occurring within its exclusive economic zone may also occur in the Arctic Ocean thus making it also an Arctic Ocean coastal state (Quinn 2015; Wegge 2015).

It has been suggested that the Oslo Declaration, like the Ilulissat Declaration before it, constitutes some sort of Arctic exceptionalism, aimed at ensuring the A5 retain the 'upper hand' in the unfolding political processes in the Arctic (Wegge 2015, p. 337). A close reading of the Oslo Declaration suggests that this may well have been its intent. However, while its political effect should not be underestimated, *as a legal matter* the Oslo Declaration does relatively little to fill the regulatory gaps relating to high seas fishing in the Central Arctic Ocean. Moreover, even if that was the intent of the Declaration, this has been rather ameliorated by subsequent events, including the convening of a 'Broader Process' as envisaged in the Oslo Declaration involving other Arctic and non-Arctic states.

This chapter examines the emerging regime for the regulation of high seas fisheries in the Arctic. In particular it interrogates the notion that in adopting the Oslo Declaration the A5 have somehow done something that is anathema to the interests of the international community in the fish resources of the Central Arctic Ocean. It begins with a brief description of the fisheries resources of the Central Arctic Ocean and the challenges that warming oceans might present for their regulation. It then describes the international legal framework for the regulation of fisheries in the Central Arctic Ocean before turning to an analysis of the Oslo Declaration and the Broader Process. It will be concluded that the Declaration itself constrains the rights and interests of neither the A5 nor the international community. Of greater import will be the outcome of the subsequent negotiations it has spawned.

The Fish Resources of the Central Arctic Ocean
and the Challenge of Warming Oceans

While many geographical definitions of the Arctic and of the Arctic Ocean exist (Rayfuse 2007, pp. 197–198), the Oslo Declaration is only concerned with 'high seas fishing in the central Arctic Ocean'. According to Molenaar, this suggests that the central Arctic Ocean consists of both high seas areas and adjacent areas (Molenaar 2016b). As a geographical matter this interpretation makes perfect sense. However, as a legal matter it is critically important to distinguish between the two areas. Thus, for present purposes, the terminology of 'Central Arctic Ocean' is used here to refer to the area of high seas covering approximately 2.8 million square kilometres which lies in the centre of the Arctic Ocean both beyond and completely surrounded by the exclusive economic zones of the A5. In other words, this chapter concerns itself only with the high seas and only with high seas fishing in the Central Arctic Ocean, and not in the other 'Arctic' high seas areas such as the 'Banana Hole' in the North East Atlantic, the Barents Sea 'Loophole', or the 'Donut Hole in the Central Bering Sea.

According to the Census of Marine Life, more than 200 species of fish are found in Arctic waters (COML 2010). However, when it comes to the fishery resources of the Central Arctic Ocean, precious little is known (Bluhm et al. 2015). Whether this is because the polar ice cap has thus far made fishing impossible or because there are simply no fish there has been a matter of conjecture, with most evidence supporting the latter conclusion (FiSCAO 2015b; Shephard et al. 2016). In recent years, however, increasing ocean temperatures coupled with decreasing sea ice coverage caused by climate change have been linked with the northward expansion of sub-arctic and temperate fish species (Wassmann et al. 2011; Christensen et al. 2014) with six stocks in particular being identified as having a 'high potential' to expand into the Arctic Ocean (Hollowed et al. 2013). Of particular interest is the polar cod (*Boreogadus saida*), the most abundant Arctic fish, first and second year juveniles of which are known to be found under the pack ice in the Eurasian basin (David et al. 2016). While nearly all such migration or expansion is expected to occur within the exclusive economic zones of the A5, the possibility clearly exists for their eventual expansion into the Central Arctic Ocean. Scientists remain sceptical of any such northwards expansion into the Central Arctic Ocean in the short term (FiSCAO 2015a, b). However, with approximately 40% of the Central Arctic Ocean, primarily in the Beaufort, Chuchki and East Siberian Seas north of Canada, Russia and the United States, now ice free in summer (Overland and Wang 2013) and warming trends set to continue, and with extensive fishing industries already operating in some Arctic and sub-Arctic regions, the potential also exists for fishing vessels to move into the Central Arctic Ocean to explore for and, if found, commercially exploit fishery resources.

Unfortunately, the history of commercial exploitation of fish stocks is replete with instances of over-exploitation and stock collapse. Particularly in situations where little is known about a species or a particular fish stock, unregulated expansion into new fisheries may effectively wipe out a species or stock before its existence is even formally recognised or understood. A classic, but by no means the only, example is the rapid expansion of the pollock fishery in the high seas of the Central Bering Sea in the 1980s, which started when the United States expelled foreign fishing fleets from its newly declared exclusive economic zone. Within less than a decade the catch taken rose from 18,000 mt in 1980 to a high of 1,448,000 mt in 1989. In 1992 the stock crashed yielding only 10,000 mt of Pollock (Rayfuse 2004, pp. 284–285). To this day, stocks have not recovered sufficiently to allow the moratorium put in place in 1993 by the CBS Convention to be lifted and, while stocks within the exclusive economic zones of Russia and the United States are generally in good shape, the high seas fishery is effectively *functus*.

The risk of stock collapse due to over-exploitation will only be exacerbated by climate change induced changes in stock composition, distribution and resilience should species with an as yet tenuous relationship with their new environment be heavily targeted by commercial fishing (Rayfuse 2012). Within the exclusive economic zone, stock migration will pose difficult issues for national fisheries management authorities seeking to prevent conflict between commercial and artisanal fishers and different gear types and to balance the need to protect vulnerable new fisheries against the desire to exploit the resource to provide income to coastal communities (Ayles et al. 2016). In the case of transboundary or shared stocks (those shared between two states), climate induced stock migration may affect the share of a stock in each state. As both states seek to maintain their share of the catch, this may have adverse implications for stock management and for stock status. If the state with the diminishing percentage of the stock fails to reduce its catch then it may undermine conservatory efforts and catch limits in the other country. In a worst case scenario, continued take by the state losing the stocks coupled with increased take by the state acquiring more of the stock could lead to the stock being fished to extinction.

The problem is even more acute in the case of high seas fisheries where a range shift away from the coastal state will weaken its conservation incentives and aggravate management as between that state and any relevant high seas fishery regime. If no high seas regime exists, new or increasing fishing pressure in high seas areas adjacent to areas under national jurisdiction may have devastating consequences for conservation and management within areas under national jurisdiction and lead to conflict between coastal states and high seas fishing states. A range shift to a coastal state will similarly aggravate management and conservation status if it leads to increased fishing pressure within areas under national jurisdiction and no corresponding reduction in the high seas area. Dramatic shifts in migration could be particularly problematic in the case of highly migratory species, such as anadromous species and tuna, in areas where pockets of high seas are interspersed with or surrounded by areas under national jurisdiction. These considerations are particularly relevant to the Central Arctic Ocean where, absent any current discrete high seas fish stocks, future fish stocks will all arrive via migration through one or more of the EEZs of one or more of the A5.

It is precisely because of the lack of scientific data and uncertainty surrounding the existence and/or potential migration of fish stocks into the Central Arctic Ocean and the effect of any fishery on what is thought to be the extremely fragile marine ecosystem of the Central Arctic Ocean that by 2007 academics and NGOs were calling for a ban on commencement of any commercial fishing in the Central Arctic Ocean pending the establishment of scientifically sound baselines and management measures (Rayfuse 2007). Amidst growing concern, in 2012 an open letter from 2000 scientists from 67 countries identified the need for biological information to understand the presence, abundance, structure, movements and health of fish stocks and the role they play in the broader ecosystem of the Central Arctic Ocean in order to enable the adoption of a robust management system and to better understand the effects of fishing removals on other components of that ecosystem (An 2012; Baker 2012). Inuit leaders, too, called for a moratorium on fishing in the Central Arctic Ocean until fish stocks have been adequately assessed and a sustainable management regime – which involves the Inuit – is in place (Kitigaaryuit Declaration 2014). As discussed in the following section, far from mere novelty, the legal basis for such measures is clearly found in the law of the sea.

The International Legal Framework for Fisheries in the Central Arctic Ocean

It is important to remember that within the exclusive economic zone (EEZ) it is the coastal state that has jurisdiction over the conservation and management of fisheries. While the LOSC provides that any 'surplus' resources not harvested by national fleets should be made available for harvest by foreign fleets, the determination of the existence of any surplus is wholly within the power of the coastal state (LOSC, Art. 62). This is particularly relevant in the Arctic Ocean where, unless and until fish stocks are found to exist in the Central Arctic Ocean, it is the A5 who will have sole jurisdiction over any fisheries located within their EEZs.

With respect to the Central Arctic Ocean, the basic legal framework governing fisheries is set out in the provisions of the LOSC dealing with the conservation and management of marine living resources in the high seas (LOSC, Arts. 116–119) and its implementing agreement on straddling and highly migratory fish stocks, the 1994 Fish Stocks Agreement (FSA), as well as a number of other instruments that have been adopted under the UN FAO such as the 1992 Compliance Agreement, the various international plans of action, and the 2009 Port State Measures Agreement and certain United Nations General Assembly Resolutions such as those relating to large scale high seas driftnet fishing and the protection of vulnerable marine ecosystems from destructive fishing practices such as bottom trawling.

As a basic proposition, all states have the right for their nationals to fish on the high seas. However, this right is subject to the duties to conserve and to cooperate with other states in the conservation and management of the resources. Importantly, the right is also subject to the rights, duties and interests of coastal states in, among other things, straddling and highly migratory species (LOSC, Art. 116).

The duty to conserve requires states to adopt appropriate conservation and management measures both individually and in cooperation with other states whose nationals exploit similar resources or different resources in the same area (LOSC, Arts. 117 and 118). These measures are to be based on the best scientific evidence available and aimed at maintaining or restoring populations of harvested species at levels which can produce the maximum sustainable yield (LOSC, Art. 119). Regular exchange of available scientific information and relevant data through appropriate international organisations is required and catch limits and conservation measures are to take into consideration the effects on dependent and associated species with a view to ensuring they, too, are maintained at or restored to levels above which their reproduction may be seriously threatened. In other words, an ecosystem approach is required. The Fish Stocks Agreement builds on this by requiring both an ecosystem and a precautionary approach to conservation and management for the purpose of ensuring long-term sustainability and promoting optimum utilisation (FSA, Arts. 5 and 6).

The duty to cooperate requires all exploiting states to agree on, and implement, measures to regulate exploitation. The recognised *modus operandi* for this cooperation is through the establishment of subregional or regional fisheries management organisations (RFMO) although it is also recognised that there may be some cases where a formal organisational structure is unnecessary and the objectives of conservation and management can be met through an 'arrangement' (RFMA) (LOSC, Arts 117, 118). The Fish Stocks Agreement further institutionalises the duty to cooperate by requiring its exercise through RFMO/As (FSA, Art. 8). States are to enter into consultation in good faith and without delay to reach agreement on arrangements, particularly where evidence exists that fish stocks may be under threat of over-exploitation or where a new fishery is being developed. Where an RFMO/A already exists coastal and fishing states are obliged either to become members of the organisation or to agree to apply its conservation and management measures and all states having a 'real interest' in the fisheries concerned may join the relevant RFMO/A. Where no RFMO/A exists, states are obliged to cooperate to establish one, or to establish other appropriate arrangements to ensure conservation and management of the stocks concerned, and to participate in the work of these organisations or arrangements. Importantly, the LOSC particularly singles out states bordering areas of high seas that are wholly or partially enclosed by the exclusive economic zones of one or more states and in which fishing activities take place, specifically requiring these states to cooperate with each other in the performance of their rights and duties under the Convention and to invite other interested states to join them in doing so (LOSC, Art 123).

In terms of the Arctic Ocean as a whole, commentators regularly refer to a range of existing RFMO/As as relevant (see, e.g., Molenaar 2013). However, when it comes to the Central Arctic Ocean the only currently relevant RFMO/As are the North East Atlantic Fisheries Commission (NEAFC) and the Joint Norwegian-Russian Fisheries Commission (Joint Commission). The geographic scope of NEAFC includes the portion of the Central Arctic Ocean between 44° West and 51° East up to the North Pole (NEAFC Convention, Art. 1). While the precise geographical area of application of the Joint Commission is unclear (Molenaar 2016a, pp. 440–445), even if, as Molenaar suggests, it can be assumed to apply to species whose distributional ranges may extend into the Central Arctic Ocean, as a bilateral arrangement between Norway and Russia it only applies to those states and their nationals.

Two other RFMOs which may become relevant in the future are the International Commission for the Conservation of Atlantic Tunas (ICCAT) and the North Atlantic Salmon Conservation Organisation (NASCO). However, the ICCAT convention area is defined as 'all waters of the Atlantic Ocean including its adjacent seas' (ICAAT Convention, Art. 1). The map of the convention area on the Commission's website indicates its area of competence as extending only to 70° North. While this map may not be determinative, the accepted official characterisation of the Arctic Ocean as an *ocean* and not a mere *sea* adjacent to the Atlantic leaves the convention's application to any Atlantic tuna or tuna-like species that may appear in the Central Arctic Ocean in serious doubt. The NASCO convention, for its part, applies to all Atlantic salmon species originating in waters north of 35° North throughout their migratory range and prohibits all fishing for salmon on the high seas (NACSO Convention, Art. 2). This prohibition, however, adds nothing to the globally applicable prohibition on high seas salmon fishing set out in Article 66 of the LOSC and any management of such species will thus be the sole reasonability of the coastal states.

In short, according to the law of the sea, the A5 are under a positive obligation to cooperate both amongst themselves and with other 'interested states' in the conservation and management of the fish stocks of the Central Arctic Ocean. At least in the case of straddling and highly migratory fish stocks – which it is expected any Central Arctic Ocean fish stocks will be – this cooperation is to be carried out either directly on a bilateral basis or through the establishment of one or more RFMO/As. The only existing RFMO/A relevant to the Central Arctic Ocean that enjoys any sort of international participation is NEAFC, whose parties are limited to Denmark (in respect of the Faroe Islands and Greenland), the EU, Iceland, Norway and Russia. As a matter of basic treaty law, NEAFC regulations do not apply to any non-member states. Moreover, the sector of the NEAFC convention area which lies in the Central Arctic Ocean represents only 8% of the total area of the Central Arctic Ocean (Pew 2012). Thus, the A5, remain under an obligation to cooperate both amongst themselves and with other 'interested states' to conserve whatever fisheries resources might exist in the remaining 92% of the Central Arctic Ocean. Of course who might constitute an 'interested state' is something of an open question given that no state has ever fished in the Central Arctic Ocean (Molenaar 2004). However, given this context, it is hardly surprising that the A5 moved first collectively to adopt the Oslo Declaration before opening the negotiating doors to others.

The Oslo Declaration on High Seas Fishing in the Central Arctic Ocean

International discussions on the future management of Central Arctic Ocean fisheries have their origins in a joint resolution passed by the United States Senate in 2007 'directing the United States to initiate international discussions and take necessary steps with other Nations to negotiate an agreement for managing migratory and transboundary fish stocks in the Arctic Ocean' (US Senate 2007). In 2008, with increasing European interest in the Arctic, the European Union proposed the expansion of NEAFC to cover the entire Central Arctic Ocean (EU 2008). However, this was rejected by some of the A5, particularly given that not all of the A5 are members of NEAFC (Molenaar 2009).

In 2009 the United States raised the possibility of convening an intergovernmental meeting aimed at adopting a non-legally binding instrument on Arctic fisheries during a side event at the meeting of the FAO Committee of Fisheries, and the question of the regulation of Arctic Ocean fisheries was broached during negotiations on the UNGA annual resolutions on Oceans and 'Sustainable Fisheries' during 2008 and 2009 (Ryder 2015). However, no tangible outcomes were produced from these discussions due to objections by at least some of the A5 to what was perceived as external meddling in their area of special interest. Nevertheless, these discussions did serve to consolidate the conviction, expressed in the Chair's Summary of the second Arctic Ocean Foreign Ministers Meeting which took place in 2010, that the A5 'have a unique interest and role to play in current and future efforts for the conservation and management of fish stocks' in the Arctic Ocean and that the development of any new international instrument on Arctic Ocean fisheries should be both initiated and led by them outside the framework of any other existing mechanisms (A5 2010). In the event, between June 2010 and July 2015 the A5 convened a number of policy and science meetings which culminated in the adoption of the Oslo Declaration on 16 July 2015 (described in Wegge 2015; Molenaar 2016b; and Shephard et al. 2016).

In the Oslo Declaration the A5 'recognise that, based on scientific information, commercial fishing in the high seas portion of the Central Arctic Ocean is unlikely to occur in the near future and, therefore, that there is no need at present to establish any additional RFMO for this area. Nevertheless, given the obligation to cooperate in the conservation and management of marine living resources in high seas areas, including the obligation to apply a precautionary approach, [they] share the view that it is desirable to adopt interim measures to deter unregulated fishing in the future in the high seas portion of the Arctic Ocean.' To that end they agree to implement measures authorizing commercial high seas fishing by their vessels 'only pursuant to any regional or subregional fisheries management organizations or arrangements that are or may be established to manage such fishing in accordance with recognized international standards'. They undertake to establish a joint program of scientific research aimed at improving the understanding of the ecosystems in the area, to cooperate with relevant scientific bodies, to promote compliance by, inter alia, coordinating monitoring, control and surveillance activities in the Central Arctic Ocean, and to ensure that non-commercial fishing is based on scientific advice, and is monitored, and that data obtained through that fishing is shared. Importantly, they specifically note that nothing in the Declaration is intended to undermine existing bodies like NEAFC, or to prejudice the rights and duties of other states pursuant to the LOSC, the FSA and other relevant international agreements.

A sanguine reading of the Declaration suggests the articulation of a precedent setting, precautionary approach to the management of fish stocks in the Central Arctic Ocean (Shephard et al. 2016). However, upon closer reading the Declaration appears to do little more than preserve the status quo. To begin with, the Declaration is only relevant and applicable to the A5. There is nothing to stop other states from engaging in the freedom to fish in the Central Arctic Ocean, as guaranteed by the LOSC, should the environmental conditions allow. Moreover, the Declaration is only applicable to the high seas of the Central Arctic Ocean. Nothing in the Declaration purports to restrict the ability of the A5 to exploit fish stocks within their EEZs, where new fishing opportunities are more likely to occur in the short to medium term and where over-exploitation will have adverse consequences for the eventual migration of stocks into the Central Arctic Ocean. Furthermore, it must be remembered that neither the Declaration nor the measures it prescribes are legally binding. The measures are thus non-enforceable even as between and against the A5. In addition, the measures only apply in respect of commercial fishing. Thus, scientific, subsistence, recreational and other non-commercial fishing activities remain permitted. As the example of Japanese 'scientific whaling' in the International Whaling Commission makes clear, exceptions in respect of scientific (and other) fishing leave open the possibility of abuse.

Even leaving aside the possibility of abuse, however, the Declaration does not, in fact, prohibit all commercial fishing. Rather, it makes it subject to measures adopted by existing and future RFMO/As and specifically stipulates that 'these interim measures will neither undermine nor conflict with the role and mandate of any existing international mechanisms relating to fisheries, including [NEAFC]'. Thus, those of the A5 who are also members of NEAFC can authorize their vessels to fish commercially in the portion of the NEAFC regulatory area that lies in the Central Arctic Ocean subject, of course, to any relevant conservation and management measures adopted by NEAFC. Similarly, while the status of the Norway – Russia Joint Commission as an RFMO/A may be subject to some doubt it is clearly an 'international mechanism relating to fisheries', the parties to which consider its geographical scope to encompass the entire Arctic Ocean and not just the Barents Sea (Molenaar 2016a). The argument can thus be made that Norway and Russia remain entitled to authorize commercial fishing by their vessels in the Central Arctic Ocean.

In short, the Declaration does very little to avert the possibility of future commercial fishing in the Central Arctic Ocean. Since neither Canada nor the United States authorize their vessels to fish on the high seas, and vessels flagged in Greenland, Norway and Russia will be regulated under either NEAFC or the Joint Commission (or both), its only effect appears to be to possibly inhibit commercial fishing by Greenlandic vessels in the non-NEAFC area of the Central Arctic Ocean. As Ryder puts it, 'at best, the Declaration can be seen as a political agreement among the [A5] to prevent *unregulated* commercial fishing by their vessels' (Ryder 2015, p. 6). This may of course be a not inconsiderable result, particularly given the possibility of operations by Russian fishing vessels in areas of the Central Arctic Ocean outside the jurisdiction of NEAFC or the Joint Commission. However, its limited significance does rather beg the question as to why the A5 bothered adopting the Declaration at all.

This question is particularly relevant given the ill-will the Declaration provoked on the part of other states. Even the limited scope of the Declaration did not stop Iceland from vociferously objecting to having been left out of the process (Quinn 2015). However, even if the distributional ranges of fish stocks that occur in the southern areas of the Arctic Ocean overlap with Iceland's maritime zones, no such stocks exist in the Central Arctic Ocean. Moreover, as a simple matter of geography, it is very clear that Iceland is not a coastal state in respect of the Central Arctic Ocean. For their part, Finland, the EU and others have criticized the 'utilization oriented' (Wegge 2015, p. 337) approach of the A5 which, rather than positively articulating a moratorium on commercial fishing, merely establishes an interim prohibition on unregulated fishing and a process to find out what is there with a view to the eventual regulation of its eventual exploitation. In this respect the Oslo Declaration adopts a fundamentally different approach to that applied in the Central Bering Sea where the annual harvest limit is to be set at zero if the biomass of pollock in the Aleutian Basin is less than a certain amount (CBS Convention, Arts III, IV and VII). As Molenaar notes, the reason for this difference probably lies in the unease of one or more of the A5 in adopting a mechanism that would give a single state the power to block commencement of commercial fishing operations or which would commit them in advance to adopting compatible measures in their own exclusive economic zone (Molenaar 2016a, 454).

Regardless of these criticisms, however, and despite the carefully crafted wording of the Declaration and the manner in which the A5 have occupied the moral and political, if not legal, high ground in respect of the regulation of high seas fishing in the Central Arctic Ocean (Wegge 2015; Shephard et al. 2016), it is clear that the Declaration represents a first step in the fulfillment by the A5 of their positive obligations of cooperation and conservation under the LOSC. Importantly the Declaration recognises the need to expand that cooperation, explicitly acknowledging the interests of other states and expressing the desire to work with other states in a 'broader process to develop measures consistent with this Declaration that would include commitments by all interested States'.

Moving Beyond the Oslo Declaration: The 'Broader Process'

The 'Broader Process' envisaged in the Oslo Declaration commenced in Washington DC in December 2015 when the United States initiated, hosted and chaired a meeting of the A5 along with delegations from China, the European Union (EU), Iceland, Japan and South Korea (colloquially referred to as the A5 + 5) (FiSCAO 2015a). Subsequent meetings have taken place in April, July and November 2016 and in March 2017 in Washington, DC, Iqaluit, Tórshavn and Reykjavik, respectively (see FiSCAO 2016b, c, d and FiSCAO 2017). These meetings have been informed by the reports of the Third and Fourth Meetings of Scientific Experts on Fish Stocks in the Central Arctic Ocean which took place in Seattle, USA, in April 2015 and in Tromsø, Norway, in September 2016 (see FiSCAO 2015b, 2016a). These scientific meetings have included participants representing the 10 governments of the A5 + 5 as well as interested organizations such as the Arctic Council, PICES and ICES (FiSCAO 2016c).

From the outset, the working assumption of the A5 + 5, based on the scientific advice received, has been that 'it is unlikely that there will be a stock or stocks in the high seas area of the central Arctic Ocean sufficient to support a sustainable commercial fishery in that area in the near future' (FiSCAO 2015a). Whether use of the word 'sustainable' refers to economic sustainability or to environmental sustainability based on an ecosystem approach to fisheries management is an open question. However, in any event, concerns regarding the rapid changes occurring in the Arctic region have given a certain impetus to the discussions and it is clear that this assumption equally implies a belief that such fishing will, eventually, occur in the future. The discussions are thus said to be based on the recognition by all participants of the need for a precautionary approach. To that end, the ongoing commitment to the broader scientific meetings and the joint program of scientific research and monitoring they are developing is indicative of a serious desire to improve scientific understanding of both the future fisheries potential and the broader ecosystem of the Central Arctic Ocean. Nevertheless, it is the adoption and implementation of both interim and permanent measures to prevent unregulated commercial high seas fishing that will be the test of their mettle. According to the Chairmans' Statements from the meetings, all delegations have committed to taking interim measures to prevent unregulated commercial fishing in the Central Arctic Ocean, to promoting the conservation and sustainable use of the living marine resources there and to safeguarding a healthy ecosystem in the Central Arctic Ocean (FiSCAO 2015a, 2016b, c). What is not yet clear is the form and content that these commitments will take.

In terms of form, three approaches have been on the table: adjusting the Oslo Declaration to adopt a broader non-binding statement; negotiating a binding international agreement that would, among other things, commit parties to essentially the same measures expressed in the Oslo declaration; or establishing one or more new RFMO/As for the area. None of these approaches has been considered to be mutually exclusive and it has been accepted that they might be combined in a 'step-by-step or evolutionary fashion' (FiSCAO 2015a). Negotiations have revolved around a continually updated draft Chairman's text which, as circulated in October 2016 prior to the Tórshavn meeting, 'was in the format of a legally binding agreement' (FiSCAO 2016d). According to the Chairman's Statement from that meeting, 'there was general belief that these discussions have the possibility of concluding successfully in the near future'. Reading between the lines, it thus appears likely that the outcome of the Broader Process will be a legally binding agreement, although as of October 2017 no such agreement has been reached (FiSCAO 2017).

In terms of the content of any new agreement, it is clear that the Broader Process is not concerned with negotiating the establishment of any new RFMO/As. Rather, the envisaged agreement will likely merely make binding the currently voluntary interim measures articulated in the Oslo Declaration. This is implicit in the Chairman's Statements from the Tórshavn and Reykjavik meetings which note that one of the key issues still under discussion is 'the conditions under which a decision might be made to commence negotiations on an agreement to establish one or more additional RFMO/As'. Other issues still under consideration include the manner in which the agreement addresses exploratory fishing, and decision-making procedures (FiSCAO 2016d and FiSCAO 2017). The first issue is critically important to the efficacy of the agreement in that consideration of exploratory fisheries as non-commercial could lead to widespread abuse by states carrying out essentially commercial fishing under the guise of exploratory operations. The second issue is relevant to questions such as when and how to move to establish an RFMO/A, or to amend the interim measures adopted in the agreement.

Regardless of whether a binding agreement is ultimately adopted, it must be remembered that any such agreement will only be binding on its parties which, at the moment would be, at most, the A5 + 5. To date, the A5 have been careful about who they have invited to the negotiating table. Participation by Iceland in its own right and Denmark, Finland and Sweden as part of the EU, has ensured that all Arctic states are represented, thereby alleviating the concerns about A5 exceptionalism. The presence of China, Japan and Korea is reflective of their status as major global distant-water fishing states.

The presence of China, in particular, is clearly intended to act both as a check on its growing Arctic aspirations as well as a lever for ensuring provision of the hardware necessary for the conduct of research in the Central Arctic Ocean. While China's claim to special status as a 'near-Arctic' state (Pan and Huntington 2016) is both geographically and legally meaningless, like all other states it does have a legitimate interest in the living marine resources of the high seas and in the freedoms of navigation and the conduct marine scientific research in areas beyond national jurisdiction. Critically, China possesses ice breakers and other materiel which scientists are anxious to access to assist in their investigations of the marine living resources of the Central Arctic Ocean (FiSCAO 2015b; Bertelsen and Gallucci 2016). Whether, as some political scientists suggest, the inclusion of China in the negotiations is indicative of a power transition occurring in the Arctic involving a challenge to US hegemony (see Bertelsen's chapter in this volume), it is clear that accommodating China as one of the A5 + 5 provides a benign space in which 'new opportunities for collaboration based on better relations and better mutual understanding' can be forged (Pan and Huntington 2016, p. 156).

Admittedly, other states may also consider they have a 'real interest' in Central Arctic Ocean fisheries. However, with the exception of the negotiations for the South

Pacific RFMO which were open to any state or entity having an interest in the fishery resources in the convention area, state practice in the negotiation of RFMO/As in the past evidences a clear trend towards limited participation. The limitation on participation in the Broader Process is thus wholly consistent with state practice and international law, although some mechanism may ultimately be needed to deal with new entrants in the event any viable fisheries are ever established in the Central Arctic Ocean (Molenaar 2016a, p. 460).

Moreover, it must be remembered that non-parties to any new binding agreement will not be bound and will, if and when physical conditions permit, enjoy an unfettered freedom to fish in the Central Arctic Ocean. Admittedly any fishing vessels will always have to pass through waters under the jurisdiction of the coastal states, however; the possibility exists that, like the fish, the scourge of illegal, unreported and unregulated (IUU Fishing), so prevalent throughout the world's oceans, will simply migrate to the Central Arctic Ocean. Incentives will therefore be necessary to ensure participation by other states in the agreement and in any subsequent RFMO/A that is negotiated. What such incentives might be is not entirely clear. The most obvious might be an expectation of the eventual allocation of fishing opportunities. Existing fisheries agreements generally allocate fishing opportunities on the basis of historic fishing practices (Rayfuse 2015). However, since no such practice exists in the Central Arctic Ocean, it is arguable that *all* states have an equal interest in the conservation and management of the marine living resources there and that the agreement and any subsequent RFMO/A should therefore be freely open to all. Provision for such broad participation may, however, have implications for the future of the agreement. On the one hand, an increase in participation by distant-water fishing states could challenge the interests of the A5, particularly where measures adopted within their EEZs might be less stringent than those applicable in the Central Arctic Ocean. On the other hand, an increase in participation by non-fishing states could strengthen the hand of the A5 in restricting future fishing opportunities in the Central Arctic Ocean. While Article 7 of the Fish Stocks Agreement requires coastal state measures and those adopted in respect of adjacent high seas areas to be compatible, no indication is given of whose measures are to be compatible with whose. It is thus an open question as to whether it will be the A5 or other states that will have the loudest voice in the regulation of any future fishery.

Conclusion

To the casual observer it might seem that there are many more pressing issues when it comes to the Arctic than expending valuable time and resources negotiating an agreement to manage an activity that has not yet commenced in respect of a resource that may not even exist. In other words, the whole process may be much ado about nothing. However, as the negotiation of the deep-seabed mining regime demonstrates, precautionary-minded international agreements are easier to reach before vested interests have become entrenched (Rayfuse 2007). Given the uncertainties and extremely limited scientific knowledge regarding existing and potential Central Arctic Ocean fisheries resources, particularly when combined with the current lack of activity in the area, a valuable opportunity exists to implement a truly precautionary approach to their future conservation and management based on sound science and modern international fisheries management principles and practises. In this respect, the Oslo Declaration might be said to represent a 'precautionary moment' in the governance of the natural resources of the Central Arctic Ocean.

However, lest we forget, the Oslo Declaration is neither binding nor does it commit the A5 to do or refrain from doing anything they are not already doing or refraining from doing. While a benign reading of the Declaration's focus on science and cooperation evidences the A5's intention to fulfil their obligation to the international community to cooperate in the conservation and management of the living resources of the Central Arctic Ocean, it is the Broader Process which holds more precautionary promise. At the time of writing, it remains to be seen whether the 'moment' will become a lasting one.

References

A5. (2010). *Chair's summary of the second Arctic Ocean foreign ministers meeting (Chelsea, Canada, March 29)*. Retrieved from http://www.mid.ru/en/foreign_policy/news/-/asset_publisher/cKNonkJE02Bw/content/id/257162

Agreement for the Implementation of the Provisions of the United Nations Convention on the Law of the Sea of 10 December 1982 relating to the Conservation and Management of Straddling Fish Stocks and Highly Migratory Fish Stocks, 4 August 1995, 2167 UNTS 3.

Agreement relating to the Implementation of Part XI of the United Nations Convention on the Law of the Sea of 10 December 1982, 28 July 1994, 1836 UNTS 3.

An. (2012). *PEW open letter from international scientists*. Retrieved from http://www.pewtrusts.org/en/projects/arctic-ocean-international/solutions/2000-scientists-urge-protection

Ayles, B., Porta, L., & Clarke, R. M. V. (2016). Development of an integrated fisheries co-management framework for new and emerging commercial fisheries in the Canadian Beaufort Sea. *Marine Policy, 72*, 246–254.

Baker, B. (2012). Scientists move to protect central Arctic fisheries. *Bioscience, 62*(9), 852.

Bertelsen, R. G., & Gallucci, V. (2016). The return of China, Post-Cold War Russia, and the Arctic: Changes on land and at sea. *Marine Policy, 72*, 240–245.

Blomfield, A. (2007). Russia claims North Pole with Arctic flag stunt. *The Telegraph*, 1 August 2007. Retrieved from http://www.telegraph.co.uk/news/worldnews/1559165/Russia-claims-North-Pole-with-Arctic-flag-stunt.html

Bluhm, B. A., Kosobokova, K. N., & Carmack, E. C. (2015). A tale of two basins: an integrated physical and biological perspective of the Deep Arctic Ocean. *Progress in Oceanography, 139*, 89–121.

Borgerson, S. (2008). Arctic Meltdown: The economic and security implications of climate change. *Foreign Affairs, 87*, 63077.

Census of Marine Life. (2010). *Arctic Ocean diversity*. Retrieved from http://www.arcodiv.org/Fish.html

Christensen, J. S., Mecklenburg, C. W., & Karamushko, O. V. (2014). Arctic marine fishes and the fisheries in light of global climate change. *Global Change Biology, 20,* 352–359.

Convention on The Convention on the Conservation and Management of the Pollock Resources in the Central Bering Sea, 16 June 1994.

Cressey, D. (2007). Russia at forefront of Arctic land-grab. *Nature 448,* 520–521. Retrieved from http://www.nature.com/nature/journal/v448/n7153/full/448520b.html

David, C., Lange, B., Krumpen, T., Schaafsma, F., van Franeker, J. A., & Flores, H. (2016). Under-ice distribution of polar cod *Boreogadas saida* in the central Arctic Ocean and their association with sea-ice habitat properties. *Polar Biology, 39,* 981–994.

ENB. (2016). *Second Session of the Preparatory Committee Established by the UN General Assembly Resolution 69/292 "Development of an International Legally Binding Instrument under the United Nations Convention on the Law of the Sea on the Conservation and Sustainable Use of Marine Biological Diversity of Areas Beyond National Jurisdiction".* New York 26 August–9 September). Retrieved from http://www.iisd.ca/oceans/bbnj/prepcom2/

EU. (2008). *Communication from the commission to the European parliament and the council on the European union and the Arctic region.* COM (2008) 763 of 23 Nov 2008.

FiSCAO. (2015a). *Chairman's statement on the first meeting on high seas fisheries in the central Arctic ocean.* Washington, DC, 1–3 December. Retrieved from http://www.afsc.noaa.gov/Arctic_fish_stocks_fourth_meeting/pdfs/Chairman's%20Statement%20from%20Washington%20Meeting%20December%202015.pdf

FiSCAO. (2015b). *Final report of the third meeting of scientific experts on fish stocks in the central Arctic ocean.* Washington, DC, 14–16 April. Retrieved from https://www.afsc.noaa.gov/Arctic_fish_stocks_third_meeting/meeting_reports/3rd_Arctic_Fish_ Final_Report_10_July_2015_final.pdf

FiSCAO. (2016a). *Chairman's statement on the fourth meeting of scientific experts on fish stocks in the Central Arctic ocean.* Tromsø, 26–28 September. Retrieved from http://www.afsc.noaa.gov/Arctic_fish_stocks_fourth_meeting/pdfs/4th_FiSCAO_Chairmans_Statement_Final.pdf

FiSCAO. (2016b). *Chairman's Statement on the Second Meeting on High Seas Fisheries in the Central Arctic Ocean.* Washington, DC, 19–21 April. Retrieved from http://www.afsc.noaa.gov/Arctic_fish_stocks_fourth_meeting/pdfs/Chairman's_Statement_from_Washington_Meeting_April_2016-2.pdf

FiSCAO. (2016c). *Chairman's Statement on the Third Meeting on High Seas Fisheries in the Central Arctic Ocean.* Iqaluit 6–8 July 2016. Retrieved from http://www.afsc.noaa.gov/Arctic_fish_stocks_fourth_meeting/pdfs/Iqaluit_Final_Chairmans_Statement_from_Iqaluit_Arctic_HS_ Meeting_July_2016.pdf

FiSCAO. (2016d). *Chairman's Statement on the Fourth Meeting on High Seas Fisheries in the Central Arctic Ocean.* Tórshavn, The Faroe Islands, 29 November – 1 December. Retrieved from http://arcticjournal.com/press-releases/2733/meeting-high-seas-fisheries-central-arctic-ocean

FiSCAO. (2017). *Chairman's Statement on the Fifth Meeting on High Seas Fisheries in the Central Arctic Ocean.* Reykjavik, Iceland, 27 March. Retrieved from http://www/state.gov/e/oes/ocns/opa/rls/269126.htm

Hollowed, A., Planque, B., & Leong, H. (2013). Potential movement of commercial fish and shellfish stocks from the sub-Arctic to the Arctic ocean. *Fisheries Oceanography, 22,* 355–370.

Ilulissat Declaration 2008. (2008, May 28). Retrieved from http://www.oceanlaw.org/downloads/arctic/Ilulissat_Declaration.pdf.

Kitigaaryuit Declaration. (2014). *Inuit circumpolar council of Canada (July 24).* Retrieved from http://www.inuitcircumpolar.com/uploads/3/0/5/4/30542564/declaration_english.pdf

Koivurova, T., & Molenaar, E. J. (2009). *International governance and regulation of the marine arctic: Overview and gap analysis.* Oslo: WWF International Arctic Programme. Retrieved from http://www.worldwildlife.org/publications/international-governance-and-regulation-of-the-marine-arctic-three-reports-prepared-for-the-wwf-international-arctic-program

Molenaar, E. J. (2004). Regional fisheries management organisations: Issues of participation, allocation and unregulated fishing. In A. G. Oude Elferink & D. R. Rothwell (Eds.), *Oceans management in the 21st century: Institutional frameworks and responses* (pp. 69–86). Leiden: Brill.

Molenaar, E. J. (2009). Arctic fisheries conservation and management: Initial steps of reform of the international legal framework. *Yearbook of Polar Law, 1*, 427–463.

Molenaar, E. J. (2013). Arctic fisheries management. In E. J. Molenaar, A. G. Oude Elferink, & D. R. Rothwell (Eds.), *The law of the sea and the polar regions: Interactions between global and regional regimes* (pp. 243–266). Leiden: Martinus Nijhoff.

Molenaar, E. J. (2016a). International regulation of Central Arctic ocean fisheries. In M. H. Nordquist, J. N. Moore, & R. Long (Eds.), *Challenges of the changing Arctic* (pp. 429–463). Leiden/Boston: Brill/Nijhoff.

Molenaar, E. J. (2016b, February 5). *The December 2015 Washington meeting on high seas fishing in the Central Arctic Ocean* [Web log comment]. Retrieved from http://site.uit.no/jclos/2016/02/05/the-december-2015-washington-meeting-on-high-seas-fishing-in-the-central-arctic-ocean/

Oslo Declaration Concerning the Prevention of Unregulated High Seas Fishing in the Central Arctic Ocean, 16 July 2015. Retrieved from https://www.regjeringen.no/globalassets/departementene/ud/vedlegg/folkerett/declaration-on-arctic-fisheries-16-july-2015.pdf

Overland, J. E., & Wang, M. (2013). When will the summer Arctic be nearly ice free? *Geophysical Research Letters, 40*(10), 2097–2101.

Pan, M., & Huntington, H. P. (2016). A precautionary approach to fisheries in the Central Arctic Ocean: Policy, science and China. *Marine Policy, 63*, 153–157.

PEW. (2012). *The international waters of the CAO: Protecting fisheries in an emerging ocean.* Retrieved from http://www.pewtrusts.org/en/research-and-analysis/issue-briefs/2013/04/the-international-waters-of-the-central-arctic-ocean-protecting-fisheries-in-an-emerging-ocean

Quinn E. (2015). *Iceland blasts Arctic five for exclusion from fishing agreement.* Retrieved from http://www.rcinet.ca/eye-on-the-arctic/2015/07/30/iceland-blasts-arctic-five-for-exclusion-from-fishing-agreement/

Rayfuse, R. (2004). *Non-flag state enforcement in high seas fisheries.* Leiden: Martinus Nijhoff.

Rayfuse, R. (2007). Melting moments: The future of polar oceans governance in a warming world. *Review of European Community and International Environmental Law, 16*(2), 196–217.

Rayfuse, R. (2012). Climate change and the law of the sea. In R. Rayfuse & S. Scott (Eds.), *International law in the era of climate change* (pp. 147–174). Cheltenham: Edward Elgar.

Rayfuse, R. (2015). Regional fisheries management organizations. In D. R. Rothwell, A. G. Oude Elferink, K. Scott, & T. Stephens (Eds.), *The Oxford handbook of the law of the sea* (pp. 439–462). Oxford: Oxford University Press.

Reynolds, P. (2007). Russia ahead in Arctic "gold rush". *BBS News*, 1 Aug 2007. Retrieved from http://news.bbc.co.uk/2/hi/in_depth/6925853.stm

Ryder, S. (2015, July 15). *The declaration concerning the prevention of unregulated high seas fishing in the Central Arctic Ocean* [We log comment]. Retrieved from http://ablawg.ca/2015/07/31/the-declaration-concerning-the-prevention-of-unregulated-high-seas-fishing-in-the-central-arctic-ocean/

Sale, R., & Potapov, E. (2010). *The scramble for the Arctic: Ownership, exploitation and conflict in the far north.* London: Francis Lincoln.

Shephard, G. E., Dalin, K., Peldszus, R., Aparício, S., Beumer, L., Birkeland, R., Gkikas, N., et al. (2016). Assessing the added value of the recent declaration on unregulated fishing for sustainable governance of the Central Arctic Ocean. *Marine Policy, 66*, 50–57.

United Nations Convention on the Law of the Sea, 10 December 1982, 1833 UNTS 3.

US Senate. (2007). Joint *Resolution No 17 of 2007, Senate 4 October 2007, House of Representatives May 2008, President Bush signature 4 June 2008*. Public law 110–243, 122 STAT. 1569–1571 (3 June 2008).

Vaughan, R. (1994). *The Arctic: A history.* Stroud: Sutton Publishing Limited.

Wassmann, P., Duarte, C. M., Agustí, S., & Sejr, M. K. (2011). Footprints of climate change in the Arctic marine ecosystem. *Global Change Biology, 17*(2), 1235–1249.

Wegge, N. (2015). The emerging politics of the Arctic Ocean: Future management of the living marine resources. *Marine Policy, 51*, 331–338.

Young, O. (2016). Governing the Arctic Ocean. *Marine Policy, 72*, 271–277.

A Half Century in the Making: Governing Commercial Fisheries Through Indigenous Marine Co-management and the Torngat Joint Fisheries Board

Jamie Snook, Ashlee Cunsolo, and Robyn Morris

Abstract A network of Indigenous co-management organizations is alive and robust within the management of fisheries in Canada and, subsequently, forms an important part of Arctic marine governance. This chapter examines Indigenous co-management in the Labrador Inuit Settlement Region of Nunatsiavut, Labrador through a case study of the Labrador Inuit Land Claims Agreement and the Torngat Joint Fisheries Board (TJFB). Through an analysis of the continuum of control of fish management policies in Nunatsiavut, and the resulting social, ecological, and economic outcomes, of Northern Shrimp, Snow Crab, and Arctic Char case studies, this chapter will illustrate the opportunity to engage the co-management organizations and processes to create more value for Inuit communities, and opportunities to facilitate further Indigenous participation in fisheries – engagement which ultimately will create healthier communities and ecosystems. In so doing, this chapter argues for a shift away from legal interpretation of the land claims documents, and calls for more emphasis to be placed on the spirit and intent of these documents in order to encourage and initiate dialogues and actions that are intended to meet and exceed the objectives of the land claims themselves.

Keywords Indigenous co-management • Land claims • Marine governance • Nunatsiavut • Torngat Joint Fisheries Board • Arctic • Reconciliation • Inuit

J. Snook (✉)
Torngat Wildlife, Plants and Fisheries Secretariat, 217 Hamilton River Road, Happy Valley-Goose Bay, Labrador, Canada

Department of Population Medicine, Ontario Veterinary College, University of Guelph, ON, Canada
e-mail: jamie.snook@torngatsecretariat.ca

A. Cunsolo
Labrador Institute of Memorial University, Happy Valley-Goose Bay, Labrador, Canada
e-mail: ashlee.cunsolo@mun.ca

R. Morris
Torngat Wildlife, Plants and Fisheries Secretariat, 217 Hamilton River Road, Happy Valley-Goose Bay, Labrador, Canada
e-mail: robyn.morris@torngatsecretariat.ca

© Springer International Publishing AG 2018
N. Vestergaard et al. (eds.), *Arctic Marine Resource Governance and Development*,
Springer Polar Sciences, https://doi.org/10.1007/978-3-319-67365-3_4

Introduction: Land Claims Based Indigenous Fisheries Co-management in Canada

Indigenous peoples have the right to participate in decision-making in matters which would affect their rights, through representatives chosen by themselves in accordance with their own procedures, as well as to maintain and develop their own indigenous decision- making institutions. *Article 18, United Nations Declaration on the Rights of Indigenous Peoples*

There is a sense of urgency in marine governance across the Arctic and Circumpolar North, with pressures from climate change (warmer ocean and surface air temperatures, shifting ocean currents, changing seasons, and sea ice patterns) (IPCC 2014), resource development (Schartup et al. 2015), extraction and exploration (McDowell and Ford 2014), fully allocated commercial fisheries with related concerns around fish stocks (Barley Kincaid and Rose 2014), and the inherent uncertainty associated with this time of rapid change (Woollett 2007; Ford et al. 2012). Compounding these pressures is a legacy of colonialism, and the resulting disempowerment and marginalization of Indigenous peoples from decision-making opportunities and governance structures across the North (Truth and Reconciliation Commission of Canada 2015), including commercial fisheries decisions and Arctic marine governance. Despite these challenges, and recognizing their inherent rights to harvest from the land for subsistence and for economic livelihoods, Indigenous peoples throughout the Arctic waters have been asserting their participation in co-management, decision-making, political structures, and commercial and subsistence fisheries.

Indigenous co-management is alive, robust, and influential within the management of fisheries and wildlife in Canada and internationally. The first land claim based co-management models emerged in 1975 through the historic James Bay and Northern Quebec Agreement between the Cree, the Inuit of Northern Quebec, and the Government of Canada (James Bay and Northern Quebec Agreement 1975). While there are many ways in which co-management is described in the literature, it is widely supported that the process of co-management challenges the ways in which boundaries are understood, power dynamics play out, and stewardship and ownership are conceptualized (Plummer and Armitage 2007).

The academic literature continues to discuss the efficacy of these co-management models (Stevenson 2006), the challenges associated with their implementation (Snook 2010), the questions around the devolution of power (Berkes 2010), and the lessons learned (Ayles et al. 2007; Dale 2009; Kendrick 2003; Kofinas et al. 2007; Nadasdy 2007; Pinkerton 1999), and many of these models have been evaluated extensively (Bickmore 2002; Hayes 2000). In many cases, co-management is often celebrated as a new way of approaching stewardship and natural resource management (Natcher et al. 2005), particularly if Indigenous rights and knowledges were central to the process. For example, White (2008) articulated that Indigenous people "can and do wield significant influence over land and wildlife decisions through the boards established under the northern comprehensive land claims" (p. 83) and that "they may be judged successful" (p. 72). Yet, others have argued that, while co-management itself is a process of problem-solving, co-learning, and power-sharing (Berkes 2009), in many cases, the Federal government remains intent on maintaining control and final decision-making power over the natural resources, leaving co-management still within the control of governments, and allowing for limited control for Indigenous self-determination through these processes (Rodon 1998).

Today, there are 26 different land claims and four self-government agreements signed in Canada, with over 100 more at various stages of negotiation (Indigenous and Northern Affairs Canada (INAC) 2015). These negotiated agreements with Indigenous peoples in Canada have led to a plethora of co-management processes and strategies, which generally includes various levels of shared responsibilities for lands, wildlife, plants, and fisheries. These co-management organizations have created a shared responsibility for the management of natural resources throughout as much as 40% of Canada's land mass (INAC 2015). For example, the Inuit based co-management regimes alone cover all of the Canadian Arctic, creating an extensive decision-making network that has impacts at the regional, national, and international levels.[1]

Co-management continues to evolve; and in its evolution, it is increasingly attracting more and more attention, and it is increasingly understood to be a mechanism that can incorporate complexities around environmental usage, ownership, and the conservation of natural resources (Plummer and Armitage 2007). While these co-management boards are responsible for and operate within particular contexts, and while the jurisdictional powers and controls may vary, if taken together, Indigenous co-management boards are impacting large geographies and decision-making at multiple levels.

Indeed, when understood as a *network* of Indigenous co-management boards, it is clear that these boards and their decisions and recommendations have wide- and far-reaching impacts. Going further, from a fisheries and Arctic marine governance perspective, these Indigenous co-management boards are important and well-established organizations that exist, are active players in marine governance, and integrate science and traditional knowledge together to create sophisticated policy analyses and recommendations that feed into multi-level and multi-sectoral dialogues and decision-making. In this light, then, understanding how Indigenous fisheries co-management boards are included, incorporated, and integrated into global Arctic marine governance is essential to the overall dialogue and framing of current and future governance strategies in the Circumpolar North.

Within this context, this chapter illustrates the development of the Torngat Joint Fisheries Board (TJFB), an Indigenous co-management fisheries board, which emerged from the Labrador Inuit Land Claim Agreement in Labrador, Canada.

---

[1] Examples of Inuit, marine, and land claim based co-management boards throughout the Canadian Arctic include the Torngat Joint Fisheries Board, the Nunavik Marine Region Wildlife Board, the Eeyou Marine Region Wildlife Board, the Nunavut Wildlife Management Board, and the Inuvialuit Fisheries Joint Management Committee.

Through a history of the first 10 years of co-management implementation, and three case studies of Northern Shrimp, Snow Crab, and Arctic Char, this chapter argues that Indigenous co-management boards should be further strengthened, supported, included, and respected in Arctic marine governance, in order to enhance global fisheries dialogues and decision-making.

Inuit Land Claims Settlement Region of Nunatsiavut, Labrador, Canada

Inuit and their ancestors have been surviving and thriving in the Arctic and sub-Arctic for thousands of years, relying on the abundant resources from the land and water for food, clothing, and wellbeing. Today, there are approximately 155,000 Inuit living in Canada, Greenland, Alaska, and Russia (Indigenous and Northern Affairs Canada 2016).

The majority of the 60,000 Inuit in Canada live in 53 remote communities in Inuit Nunangat (Inuit Homelands), which encompasses 35% of Canada's landmass and 50% of its coastline (Inuit Tapiriit Kanatami 2016). There are four regions of Inuit Nunangat (west to east): the Inuvialuit Settlement Region (Yukon and the Northwest Territories); Nunavut; Nunavik (Northern Quebec); and Nunatsiavut (Northern Labrador). Inuit continue to rely on the land for sustenance, livelihoods, culture, and wellbeing; subsequently, decisions that are made about the ways in which wildlife and fisheries are managed and governed in these regions have direct impacts on individuals and communities (Fig. 1).

The Inuit Land Claims Settlement Area of Nunatsiavut, Labrador is home to 5.35% of the Inuit population in Canada (population: 2325) (Indigenous and Northern Affairs Canada 2016). Nunatsiavutimmuit (Inuit from Nunatsiavut) primarily live in five small coastal communities (from north to south): Nain, Hopedale, Postville, Makkovik, and Rigolet. Additionally, Nunatsiavut Government beneficiaries live within the Lake Melville communities of North West River, Mud Lake, and Happy Valley-Goose Bay, and throughout other areas of Canada.

Nunatsiavut is located at the Northern range of the Boreal forest, and has approximately 15,000 km of coastline along the Labrador Sea. Nunatsiavutimmuit continue to actively hunt, harvest, fish, and forage for wild berries and medicinal plants, a variety of fish species, land mammals such as moose, caribou, and black bear, and marine mammals such as polar bear, seal, and porpoise.

Fish resources and marine areas are deeply connected to Inuit life in Nunatsiavut during all seasons, and are essential for economic, social, and cultural wellness, as well as Inuit identity and traditional ecological knowledge. Subsistence fishing happens in all of the Nunatsiavut communities for species such as Arctic Char and Salmon. Prior to the creation of Nunatsiavut, commercial fishing in Northern Labrador was heavily reliant on Salmon and Northern Cod (May 1966). In the wake of the decline of Northern Cod resources, commercial moratoriums for these species

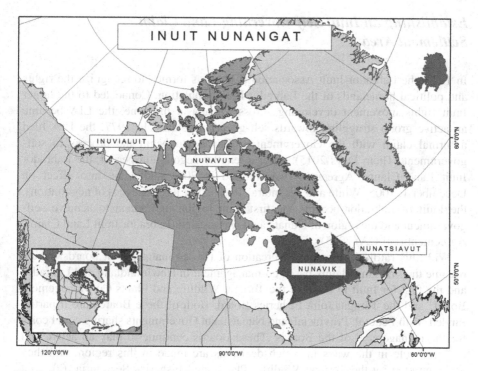

Fig. 1 Map of Inuit Nunangat. The darker the colour, the more recent the land claims settlement area (© Torngat Wildlife, Plants, and Fishers Secretariat)

were implemented in the 1990s, and the Northern Cod industry dissolved (Coombs et al. 2011).[2] Today commercial fisheries happen in Nain (char) and Makkovik (Snow Crab, shrimp, and turbot). These commercial fisheries are mainly facilitated through the Torngat Fish Producers Co-op, which formed in the early seventies when many other organizations were also mobilizing (Snook 2005). The co-op is best described here:

> [The] Torngat Co-op is the first stirring of an oppressed people that have been victimized for generations by outsiders. It is the people deriving the utmost for themselves from their own resources and it is a marching ahead of the total North coast of Labrador in a combined effort for recognition of its needs and showing Canada that we are moving ahead with confidence in ourselves and hope for ourselves. We will stumble, we will falter, and we will err, but by God, we will keep on trying. (Rennie 1989)

---

[2]Northern Cod, for example, was harvested inshore by Labrador Inuit along the entire coast, both for subsistence and for commercial livelihoods. Interestingly, because the Northern Cod moratorium was in place while the Nunatsiavut Land Claims Settlement process was being negotiated, Cod was not included within the Labrador Inuit Land Claim Agreement. This means that should the Northern Cod fishery return, there is uncertainty around how the Land Claim Agreement will apply.

Establishing an Inuit Self-Governing Land Claim Settlement Area

In 1973, the Labrador Inuit Association (LIA) was formed to recognize the rights and political homelands of the Labrador Inuit population. Connected to the larger Inuit rights movement developing across Canada at the time, the LIA became an active group struggling towards self-governing status. In 1977, the LIA filed a formal claim with the Government of Canada to begin the process to self-government (Them Days 2015). After a lengthy negotiation period, the Labrador Inuit Land Claims Agreement was ratified in 2004 and came into effect on December 1, 2005 (White and Alcantara 2016). After three decades of negotiations, the Inuit of Labrador became the first Inuit region in Canada to achieve self-government and the region of Nunatsiavut was created (Labrador Inuit Land Claims Agreement (LILCA) 2005).

With this ratification came the creation of two co-management boards, which became the first land claims-based co-management in Newfoundland and Labrador and the first for Inuit in Canada: the Torngat Wildlife and Plants Co-Management Board[3] and the Torngat Joint Fisheries Board. Both of these Boards are tripartite funded, with Federal, Provincial, and Nunatsiavut Governments sharing equal costs for the management of the Boards. These Boards continue to play an active and essential role in the ways in which decisions are made in this region, and they are supported by the Torngat Wildlife, Plants, and Fisheries Secretariat (Torngat Secretariat), which is a research and policy organization located in Happy Valley-Goose Bay (www.torngatsecretariat.ca) (Fig. 2).

The Torngat Joint Fisheries Board and Fisheries Co-management in Newfoundland and Labrador

The Torngat Joint Fisheries Board (TJFB) is the managing board for marine-based decision-making in the Labrador Inuit Settlement Area. The TJFB is actively immersed in the implementation of Indigenous co-management of marine resources in Labrador, and participates in discussions at the regional, national, and international levels. Within this light, the TJFB can be understood as an example of the strength and importance of Indigenous co-management of marine resources in the Arctic, and provides useful examples for the creation, implementation, and operations of these types of co-management boards.

---

[3]While the Torngat Wildlife and Plants Co-Management Board is beyond the scope of this paper, for more information please visit: http://www.torngatsecretariat.ca/home/torngat-wildlife-and-plants-co-management-board.htm

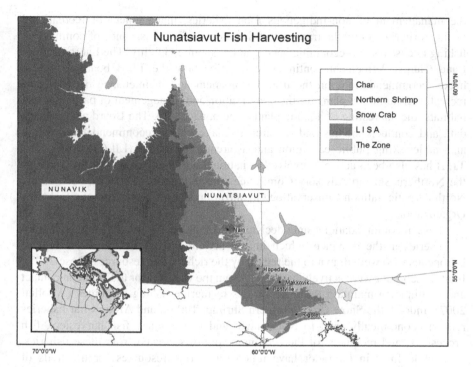

Fig. 2 Map of the communities of Nunatsiavut, Labrador, Canada, and the fishing locations and jurisdictional boundaries (© Torngat Wildlife, Plants, and Fisheries Secretariat)

The TJFB is comprised of one appointee from the Government of Newfoundland and Labrador, two from the Government of Canada, and three from the Nunatsiavut Government. The Board is overseen by an independent chairperson, who was recommended by the original six board members, who has been the first chairperson of the TJFB, and has served for the first decade of its inception and implementation. While board members are appointed by a particular Government, they are intended to be independent of the appointing Government. The primary responsibilities of the TJFB are "to make recommendations in relation to the conservation of species, stocks of fish, aquatic plants, fish habitat, and the management of fisheries in the Labrador Inuit Settlement Area" (LILCA 2005). These recommendations are targeted at the fisheries, both within the Labrador Inuit Settlement Area (LISA) and in the adjacent waters, and operate within multiple jurisdictional boundaries at the provincial, federal, and international levels.

While the primary responsibilities of the TJFB are to make recommendations in relation to the conservation of species, stocks of fish, aquatic plants, fish habitat, and the management of fisheries in the LISA, there is a wide scope of recommendation possibilities for the Board. For example, recommendations may include topics such as the waters within which harvesting may occur, the establishment of total allowable catches, allocations to the LISA, catch controls, fishing effort controls, the management of non-Indigenous food fisheries, management of recreational fishing, criteria around the transport of fish, criteria for the issuance of commercial fishing licenses, socio-economic needs, and economic viability. The Labrador Inuit Land Claims Agreement continues to describe how the TJFB board may also make recommendations on the use, management and maintenance of fishing or recreational harbours, plans for the conservation and management of particular fish habitats, the harvesting of aquatic plants, and aquaculture. The Board also collects data and conducts studies and research to make policy recommendations, while also undertaking public education and awareness programs (LILCA 2005). The TJFB has also been actively involved in national decision-making forums, such as the Northern Shrimp Advisory Committee, and international forums, such as the North Atlantic Salmon Conservation Organization and the North Atlantic Fisheries Organization.

These recommendations and decisions take place within a complex history of fisheries in the province, which impact actions today.[4] Before the advent of Europeans who were drawn to the region by the rich marine resources (Rose 2007), Inuit in Labrador were actively harvesting from the waters prior to European contact and continue to maintain these strong ties to marine fishing resources (Woollett 2007). Indeed, the Snow Crab, Northern Shrimp, Turbot, and Arctic Char fisheries remain economically vital to Nunatsiavut and to the many fish harvesters, fish processors, and recipients of the economic spin-offs created from these fisheries. Yet, while Inuit in Labrador have relied on marine resources for hundreds of years, they have been systematically marginalized in the fishing economy. The implementation of the Nunatsiavut Government, and the TJFB, however, has begun to shift this inequity and Inuit are re-entering into the commercial fishing industry in new ways, with the Nunatsiavut government successfully advocating for larger allocations of fishing resources that are Indigenous-led, Indigenous-managed, and have economic benefits for the region. Despite this advocacy, however, the TJFB and the Nunatsiavut Government are still advocating from a position of marginalization, and are still facing many hurdles, barriers, and pressures from provincial, federal, and international commercial fishing interests and regulatory bodies.

Implementation and Transition in Fisheries Co-management in Nunatsiavut

As with any establishment of a new organization, the early years of the Torngat Joint Fisheries Board were dedicated to a variety of administrative and organizational activities, such as the appointment of board members, recruiting staff, retaining staff,

---

[4]Today, various fisheries remain valuable in the Province, with an export value of almost $1 billion, and employment levels at nearly 18,000 people working in various aspects of the fishery (Community Foundation of Newfoundland and Labrador 2015).

and establishing operational procedures, governance bylaws, employee guidelines, and reporting structures. After 5 years of implementation, the Torngat Secretariat initiated research into the implementation process, to determine the strengths, challenges, and opportunities of the early years of fisheries co-management in the region, and to vision for the next 5–10 years (Snook 2010). This research involved in-depth interviews with past and present board members of the TJFB to discover from their first-hand, lived experiences what it was like to vision, implement, and develop an Inuit fisheries co-management board emergent from a land claims agreement.

This research identified that, from the Board members' perspectives, the first few years held many frustrations from system delays, including board member appointments, and some of the early administrative and human resources required to make the TJFB operational. After the first development years, however, the TJFB members saw their activities and time shifting from start-up to action and recommendations, and began to see impacts of their recommendations in practice (Snook 2010). This reflects what Berkes (2009) has argued: it often takes up to a decade before these organizations are able to solidify and begin to fulfill their mandates in impactful ways.

This research also highlighted the ways in which dialogue, influenced by multiple ways of knowing and multiple perspectives, was essential to the ways in which board members made decisions and put recommendations forward. Indeed, for many of the TJFB members, the inclusion of knowledge from lived experience, Indigenous knowledge, policy, science, research, and government created a diverse and fecund environment that fostered dialogue and created in-depth discussions – all of which was indicated to lead to the creation of stronger, more nuanced, and more accurate recommendations than would be possible without such dialogical spaces (Snook 2010).

Interestingly, one of the main challenges identified by the participants was that of "newness": newness of the region of Nunatsiavut, newness of the self-government structure, newness of the Board itself, and newness of how relationships would be formed with the provincial and federal bodies. In the early years, the participants identified that no one knew who the TJFB was nor what they were meant to do, or how they fit within jurisdictional boundaries (Snook 2010). Furthermore, and following Rodon (1998), since co-management was seen to be still lacking power or full respect in decision-making, there was a sense that government and policy makers didn't truly value or understand the development of the TJFB or its importance in marine fisheries governance.

This research also highlighted that in the case of the TJFB, governance through co-management is a process, and is premised on adaptive management and continually learning and re-shaping as new contexts, new experiences, and new information arises (c.f. Berkes 2009). Indeed, as the TJFB has developed and matured over the years, the organization has increasingly become a self-reflective entity, incorporating joint learning for the co-production of knowledge and for stronger decision-making and recommendation impacts. Interestingly, all the recommendations from

Snook's (2010) research were incorporated, further illustrating not only the adaptive learning process of the TJFB, but also the nimbleness and flexibility of organizations like the TJFB to be able to conduct and act upon their own research, as they are often unburdened by more cumbersome bureaucratic structures and histories.

Co-management in Action: Profiles on Northern Shrimp, Snow Crab, and Arctic Char

With a decade of land claim implementation experience, and 44 years since the Labrador Inuit Association was initially formed in 1973, the TJFB has emerged and matured into an entity that can contribute strongly to Arctic marine governance. In order to demonstrate the ways in which Indigenous co-management influences and impacts decision-making and marine governance in Canada and throughout the Circumpolar North, this chapter will examine three case studies which exist on a continuum of co-management power sharing (least power to most power): Northern Shrimp, Snow Crab, and Arctic Char.

The following case studies will highlight the ways in which the TJFB has impacted Arctic marine governance at the local, provincial, national, and international scales, and illustrate the importance of commercial fisheries to Inuit communities in Labrador, how the TJFB prioritized its activities based on community feedback and priorities, and how recommendations have been influenced by dialogue with all the stakeholders.

Northern Shrimp

Northern Shrimp (*Pandalus borealis*) is a keystone species that makes the commercial fishery in Nunatsiavut possible. Access to the Northern Shrimp fishery has provided individuals with employment opportunities on offshore vessels participating in the fishery. More importantly, however, cross-subsidization from Northern Shrimp allocations have allowed for continual operations of fish plants within Nunatsiavut, resulting in increased processing capabilities for Snow Crab and Turbot, along with associated employment (Coombs et al. 2010).

In the 1970s, the Federal Department of Fisheries and Oceans Canada (DFO) allocated offshore shrimp licenses to the Torngat Fish Producers Co-operative Society Ltd. and Pikalujak Fisheries Ltd., both of which are Indigenous organizations. The Nunatsiavut Government was excluded from these original allocations and, it was only later in 1997, when the Nunatsiavut Government itself finally gained access – albeit limited – to the Northern Shrimp resource through special allocations (DFO 2007).[5]

Prior to 2016, access to the Northern Shrimp fishery was based on the long-standing DFO 'Last-in, First-out' (LIFO) policy (DFO 2007). The LIFO policy was first introduced in the 2003 Integrated Fisheries Management Plan, without substantive stakeholder consultation, as a mechanism to address declines in the Northern Shrimp resource, whereby those entering the fishery last would be the first to be removed. This put the Nunatsiavut Government in an unjust position, as they only entered the fishery in the late 1990s, and actually lost quota in 2014 due to this policy (Ministerial Advisory Panel 2016). This quota loss due to the implementation of the LIFO policy was viewed as contradictory by the TJFB, when compared to other Federal initiatives. As was recommended in April 2014 by the TJFB to the Minister of DFO:

> Whereas allocation criteria were initially developed in 1997 and further developed in 2003; and having noted that the LIFO policy is absent from both, and was never the subject of any significant consultation; and recognizing Federal initiatives to increase Aboriginal access to the commercial fishery specifically, and Aboriginal economic opportunities generally; and considering LIFO to be in conflict with these Federal initiatives; the Board recommends that Aboriginal participants be exempted from the application of LIFO, and that the 1997 and 2003 access criteria form the basis for allocations through periods of decline. (Torngat Joint Fisheries Board 2014a)

In 2016, Fisheries and Oceans Canada abolished the LIFO policy for access to the Northern Shrimp fishery after an extensive Ministerial Advisory Panel process. During consultations by the Panel, the TJFB provided recommendations on the modifications of the LIFO policy. The TJFB recommended that Indigenous groups with Land Claims Agreements, particularly the Nunatsiavut Government, should be exempt from LIFO due to DFO's obligations to consider Indigenous and Treaty rights, adjacency, historical dependence, and economic viability. The final decision by the Minister included some of the key arguments made by the TJFB. The LIFO policy was replaced by a new allocation sharing regime and, as a result, the Nunatsiavut Government and other Indigenous groups in Labrador (such as the Innu Nation and the NunatuKavut Community Council) have increased their access to this fishery (Office of the Minister 2016; Ministerial Advisory Panel 2016).

Labrador Inuit, and the Nunatsiavut Government, are the most adjacent user of the Northern Shrimp resource, and although the resource is found within and adjacent to their territory, Nunatsiavut has received an inequitable and small percentage of the Northern Shrimp resource; furthermore, this quota was permitted only after the resource expanded and DFO increased allocations. Moreover, the structure and implementation of the LIFO policy created persistent uncertainty for the Nunatsiavut Government for nearly 20 years. Although the LIFO policy has

---

[5]The Nunatsiavut Government gained access to Northern Shrimp in Shrimp Fishing Areas (SFA) 4 and 5. The Zone makes up 23.9% of SFA 4 and 33.7% of SFA 5, but the Nunatsiavut Government only holds 5.1% of the quota in SFA 4, and 9.9% in SFA 5.

been abolished, and the Nunatsiavut Government's access to the shrimp resource has improved, there are still opportunities to maximize their participation in this fishery. Consequently, it is imperative that the TJFB be engaged, and thoroughly considered, in dialogue on Northern Shrimp management at the local, national, and international levels.

Snow Crab

Another important commercial species to Labrador Inuit is Snow Crab (*Chionoecetes opilio*). Labrador Inuit harvest Snow Crab throughout Nunatsiavut and these crab are processed within the Inuit community of Makkovik, which has become the centre of Snow Crab operations in Nunatsiavut. The commercial fishery for Snow Crab has emerged over the past couple of decades, and has gained significant importance for Nunatsiavut in light of the decline in commercial salmon and cod fisheries (Coombs et al. 2011).

The amount of Snow Crab that is allowed to be harvested is determined by DFO. DFO provides the Nunatsiavut Government with a communal allocation to manage locally. Since 2013, this amount has been 310 metric tonnes of Snow Crab annually. The Nunatsiavut Government designates Labrador Inuit to harvest its quota. In 2016, the Nunatsiavut Government designated seven vessels to fish its allocation, and 84% of the allocation was harvested, valued at 1.48 million dollars (TFPC 2016). DFO is responsible for scientific surveys to assess the health of the Snow Crab resource off the coasts of Newfoundland and Labrador, but in recent years, the TJFB has helped to prioritize research adjacent to the Labrador Inuit Settlement Area, and led new research in collaboration with DFO, the Nunatsiavut Government, and Snow Crab harvesters to collect the data needed to support decision-making decisions for the Nunatsiavut's Snow Crab fishery (Norsworthy et al. 2012).

The status of the resource has continually worsened in the last several years, with the stock of Snow Crab declining since the early 2000s, and the outlook on the stock is uncertain. This is partly due to a warming oceanographic regime, which is diminishing the reproductive productivity of Snow Crab (DFO 2016; Colbourne et al. 2016) and becoming more favorable for groundfish, which predate on crab. Due to this regime shift, there is some uncertainty in the future of the Snow Crab resource, and therefore, the future of the Makkovik processing plant.

The Torngat Joint Fisheries Board considered the science and future trends, Snow Crab fishery performance, and community feedback, in preparation for developing recommendations for the Minister of the DFO. In 2014, the TJFB recommended that the allocation of Snow Crab to the Nunatsiavut Government be reduced (Torngat Joint Fisheries Board 2014b). Although the Minister decided not to reduce allocation, through discussions with the Board, the Nunatsiavut Government recognized the vulnerability of the fishery and voluntarily held back 15% of their communal quota. The Nunatsiavut Government has also withheld an additional 100 metric tonnes of exploratory quota that could have been harvested.

This voluntary reduction of quota illustrates what happens when there is local control over resources that need to be sustained, and highlights how local concerns and knowledge, coupled with research, result in stewardship decision-making and precautionary management actions.

Arctic Char

The Arctic Char (*Salvelinus alpinus*) fishery in Nunatsiavut is a highly-localized fishing industry within Nunatsiavut run by and for the Labrador Inuit (Michaud et al. 2010). It has been an important subsistence resource for hundreds of years and more recently a commercial resource (Labrador Inuit Association 1977; Williamson 1997; Furgal et al. 2016). It is a wild salmonid resource that has been intentionally under harvested from the set quota level in the region in the absence of scientific certainty of the resource. In addition to economic benefits, there are also cultural and health benefits from this fishery for the Labrador Inuit. Not only does eating Arctic Char provide a valuable nutritional food source, but it also connects people to physical activity and time on the land, while simultaneously connecting to long-standing cultural practices in the region (Torngat Joint Fisheries Board 2011). As such, the TJFB has recognized the importance of Arctic Char to Labrador Inuit and have facilitated several co-management initiatives to enhance discussion of this species and to support the industry.

Yet, despite this long history of reliance and harvesting on this resource, and although Labrador Inuit can harvest to the level of their food and cultural needs, the DFO still manages the commercial fishery within Labrador (DFO n.d.). The amount of fish that can be harvested within Nunatsiavut is determined by DFO, who provides the Nunatsiavut Government with a communal license. The Nunatsiavut Government then designates Inuit fishers to harvest the quota, which can be harvested entirely within the LISA. Within the last 20 years, only approximately 30% of the quota has been harvested. This is partly due to uncertainty in the status of the resource. In the absence of counting facilities and other estimates of abundance, the status of the resource remains unknown. The last stock status, based primarily on catch rates, on Labrador Arctic Char was analyzed in 2001, and within the last 15 years, uncertainty increases.

Nevertheless, the Arctic Char fishery in Nunatsiavut represents a true grassroots industry: Arctic Char is entirely harvested for Inuit, by Inuit. All commercial Arctic Char is caught by Inuit designates, it is processed in Nain by the Torngat Fish Producers Co-operative Society Ltd., and it is sold locally in Nain and in Happy Valley-Goose Bay. Due to the direct contact and participation in all aspects of harvesting and processing, Arctic Char are highly valued in Nunatsiavut. It is an iconic species with the potential to boost the Nunatsiavut economy when commercial demand increases for Arctic Char. When this happens, it is essential that DFO include the TJFB as a key-stakeholder in any decision-making decisions, as this is the most Inuit-controlled fishery that operates entirely within the LISA.

Case Studies Summary

Each of these case studies highlight, in various manners, three key points: (1) they illustrate that the Indigenous commercial fisheries are alive and well in Nunatsiavut, and have been resilient in spite of external control and management that left Inuit out of the decision-making process; (2) they highlight the inconsistencies in management strategies for each of the species, and clearly show that the further offshore the fisheries stock, the less equitable access the Inuit have (i.e. the closer to shore the stock, the more Inuit control is maintained); and (3) they exemplify three fisheries within an Arctic marine environment that exist with high levels of uncertainty due to small amounts of budget, lack of science about the status of these fisheries, and inconsistencies in oversight and management.

All of these fisheries represent outcomes from the 'era of management' (Ludwig 2001, cited in Armitage et al. 2012) – that is, they all emerged from a system of management that was imposed from the top-down onto the co-management boards and the communities that they serve. Indeed, there is such a gap between the current state of these fisheries, and what is possible with new approaches, that there is an immense opportunity to provide more economic and social value to these Indigenous communities and to the fisheries themselves.

Given the resilience these commercial fisheries, with the right combination of strategies, policies, and will, there is an opportunity to strengthen and grow these Indigenous commercial fisheries to the point that the social and economic benefits in this region would be impactful and positive. Indigenous co-management boards, like the Torngat Joint Fisheries Board, have the ability – and indeed, the mandate – to promote healthy communities and healthy fisheries, and understand the local contexts, environment, needs, and priorities in a deeply intimate manner, drawing on multiple ways of knowing and multiple forms of sciences. These case studies make clear that Indigenous co-management boards have an important role in the global fisheries and marine governance arena, and need to be fully included, incorporated, and supported, building on the already-existing infrastructure and processes across the North.

Discussion & Moving Forward: Strengthening & Supporting Indigenous Co-management in Arctic Marine Governance

It is clear from the three case studies presented in this chapter, as well as the history and development of the Torngat Joint Fisheries Board, that Nunatsiavut has a fishing base that is strong. Yet, these gains have not come easily; the Labrador Inuit have experienced systematic marginalization and exclusion at all levels of decision-making and, indeed, were often left out of discussions and the allocation of commercial fisheries licenses occurring at the federal level.

Yet, these Indigenous co-management boards do make important contributions and impacts. The Torngat Joint Fisheries Board can and should be understood as part of a larger continuum of the assertion of power for access to marine resources and commercial opportunities, which has been almost 50 years in the making, started with the vision of a land claim and Inuit self-determination, and has emerged, over 12 years after the establishment of the Nunatsiavut Government, as an active and impactful organization that is continuing to grow, vision, and dream for increasing engagement in the national and international realms. Organizations such as the Torngat Joint Fisheries Board need to be further strengthened and supported in their mandate to mobilize co-management for stronger, healthier more vibrant people, communities, environments, and resources.

When taken alone, Indigenous co-management boards can be understood not only as the legacy of decades of political advocacy and the assertion and recognition of Indigenous rights and sovereignty, but also as active, effective, and important voices and players in Arctic marine governance and the stewardship of natural resources in Canada and throughout the Circumpolar world. The Arctic is already under a system of Indigenous co-management governance that needs to be respected and taken seriously and, in so doing, create new ways of engaging, new forms of decision-making, and a further empowering of the organizations that already exist. Within this context, then, we advocate for more attention and support to be given to the co-management boards across the Arctic, for including them more fully and authentically in dialogues and decision-making, and for supporting and empowering these organizations to continue to do their work.

Furthermore, if these co-management boards are understood as an active network, it begins to reframe conceptualizations of Arctic marine governance and resource stewardship. These co-management boards, and the contexts from which they emerged, create an important and active structure of governance across the Canadian Arctic – one that is locally-contextualized and incorporates and integrates Indigenous knowledges and sciences with other forms of science and research. By understanding co-management boards across the Arctic as a network, rather than independent organizations, it begins to reframe the discussion to include Inuit and other Arctic Indigenous peoples as active and important players in marine governance, with both commercial and subsistence interests.

Yet, there are many systematic barriers when it comes to incorporating Indigenous co-management into larger dialogues and processes. In many cases, the large state actors and those responsible for making the decisions may be far-removed from and have infrequent contact with the Indigenous co-management boards (Plummer and Fennell 2007). Indeed, according to Plummer and Fennell (2007), and following a reciprocal altruism perspective, the less frequent the interactions are between state institutions and co-management institutions, the more likely the state will forget about or ignore Indigenous rights and sovereignty. Co-management boards and state actors, then, must ensure continual opportunities for contact, discussion, and dialogues to co-produce knowledge and to continue to find ways to bring all voices to the decision-making table.

In this light, Indigenous co-management boards can be understood to emphasize the need for a paradigm shift away from legal interpretation of the land claims documents, and more emphasis in the spirit and intent of these documents as living and breathing entities that encourages the initiation of dialogues and actions that are intended to meet the objectives of the land claims themselves. The focus should not be placed on meeting the minimum legal requirements laid out in the original land claims documents; rather, these documents can be viewed as the minimum baseline from which to build all future decisions and actions, and should be continually exceeded and improved to support the health of environment and of people.

In order to have those dialogues to effect decision-making, however, they need to be informed by new and sustained science in the North. In many cases, there is limited longitudinal data on the health and stock assessments for many species in Arctic waters. This lack of science creates major challenges for evidence-based decision-making for Indigenous co-management boards and, indeed, for all levels of governments. This science vacuum can also create conditions in which federal governments or the co-management boards are not leading the science because of lack of resources allocated or internal capacities; rather, science is put in the hands of the larger off-shore fishing industry, creating tensions in resource allocations, and concerns over the reliability of the science and who is making the decisions.

On May 10, 2016, and after a decade since its inception, Canada formally signed on as a full supporter, without qualification, to the United Nations Declaration of the Rights of Indigenous Peoples (United Nations 2008). Many of the Articles in this declaration relate directly to Indigenous co-management structures and assert the importance of Indigenous self-determination and sovereignty over their lands and the natural resources within, both from a subsistence and a commercial perspective (e.g. Articles 3, 4, 5, 11, 18, 19, 20, 23, 26, 27, 29, 32). The articles cite that "Indigenous peoples have the right to participate in decision-making in matters which would affect their rights... as well as to maintain and develop their own Indigenous decision-making institutions" (18) and state that "Indigenous peoples have the right to the lands, territories, and resources which they have traditionally owned, occupied, or other-wise used or acquired... States shall give legal recognition and protection to these lands, territories, and resources" (26). In addition, emphasis is placed on the rights of Indigenous people "to determine and develop priorities and strategies for the development or use of their territories or other resources" (32), calling on states to "consult and cooperate in good faith" with Indigenous peoples and their organizations, while providing "effective mechanisms for just and fair redress for any such activities, and appropriate measures shall be taken to mitigate adverse environmental, economic, social, cultural or spiritual impact" (32).

When understood within this larger international context of Indigenous rights and sovereignty, Indigenous co-management is an already-existing mechanism that meets many of the articles and calls for action within the United National Declaration on the Rights of Indigenous Peoples (UNDRIP). Indeed, within the current era of reconciliation and Nation-to-Nation relationships in Canada, and with commitments to international agreements such as UNDRIP, Canada should be celebrating and supporting Indigenous co-management boards across the country and looking to these boards as one such way to operationalize Indigenous rights, sovereignty, and reconciliation through already-existing structures and frameworks. From an Arctic marine resource perspective, then, Indigenous co-management boards are an important component of the international call for the rights of Indigenous peoples, and no marine governance or usage policy should be made in the absence of the input from these various Indigenous organizations. If international actors and governments are serious about moving forward with Indigenous rights, then Indigenous co-management boards and the network of these boards, needs to be strengthened, respected, and incorporated into the global decision-making framework.

Conclusion: "The Era of Management Is Over"[6]

It is clear that the vision for shared decision-making and power for stewardship of natural resources that co-management represents has continued to grow in depth, scope, and influence. Each Indigenous co-management board in Canada brings nuance and local contexts to the discussions and the debate, and adds richness to the ways in which co-management processes and organizations can impact, mobilize, and transform stewardship policies.

Indigenous co-management boards can also be understood as contributing to the reconciliation process in Canada; indeed, through their resistance of top-down decision-making and the homogeneity of ontologies, epistemologies, and method-ologies (Natcher and Hickey 2002), and their increasing emphasis on the resurgence and reclamation of Indigenous leadership, control, and self-determination, Indige-nous co-management boards are re-shaping what it means to bring together multiple stakeholders, voices, and ways of knowing for conservation and stewardship.

From an Arctic marine governance perspective, then, Indigenous co-management organizations, such as the Torngat Joint Fisheries Board, are challenging the 'era of management' (Ludwig 2001; Armitage et al. 2012) of natural resources, and showing what is possible when dialogue, co-learning, cooperation, and adaptive behaviours (Berkes 2009) guide how we understand and govern marine resources in the North. Indeed, to truly move forward in Arctic marine governance and decision-making, Indigenous co-management boards need to be acknowledged, supported, and strengthened, and their recommendations and decisions respected, incorporated, and amplified. This is a network and an approach that Indigenous peoples have chosen, through the implementation of land claims agreements, which is already in existence. It can no longer be business as usual in Arctic marine governance;

---

[6]Ludwig 2001, cited in Armitage et al. 2012.

international and federal regulatory bodies need to find ways to devolve power to these Indigenous co-management boards, and to work with them in shared partnership that allows all levels of decision-making and multiple voices to work together for stronger, more robust, more sustainable outcomes during times of rapid change and complexity (c.f. Armitage et al. 2012).

Acknowledgements This chapter would not have been possible without the support of the Torngat Joint Fisheries Board (Chesley Andersen, David Bonnell, John Mercer, Derrick Pottle, Craig Taylor, and Keith Watts), past board members (Eric Andersen, Joey Angnatok, Ricky Edmunds, Stanley Oliver, Alphonsus Pittman, and Alex Saunders), colleagues at the Torngat Wildlife, Plants and Fisheries Secretariat (Rosamond Andersen, Aaron Dale, Victoria Neville, Beverly White, and Bryn Wood), and past Torngat Secretariat colleagues (Julianna Coffey and Julie Whalen). Thanks also to the University of Guelph Department of Population Medicine. The work of the Torngat Joint Fisheries Board is tripartite funded by the Nunatsiavut Government, Government of Canada and the Government of Newfoundland and Labrador. Maps and figures created by Bryn Wood of the Torngat Secretariat.

References

Armitage, D., de Loë, R., & Plummer, R. (2012). Environmental governance and its implications for conservation practice. *Conservation Letters, 5*(4), 245–255. https://doi.org/10.1111/j.1755-263X.2012.00238.x.

Ayles, B., Bell, R., & Hoyt, A. (2007). Adaptive fisheries co-management in the Western Canadian Arctic. In D. Armitage, F. Berkes, & N. Doubleday (Eds.), *Adaptive co-management: collaboration, learning and multi-level governance* (pp. 125–150). Vancouver: UBC Press.

Barley Kincaid, K., & Rose, G. A. (2014). Why fishers want a closed area in their fishing grounds: Exploring perceptions and attitudes to sustainable fisheries and conservation 10 years post closure in Labrador, Canada. *Marine Policy, 46*, 84–90. https://doi.org/10.1016/j.marpol.2014.01.007.

Berkes, F. (2009). Evolution of co-management: Role of knowledge generation, bridging organizations and social learning. *Journal of Environmental Management, 90*(5), 1692–1702. https://doi.org/10.1016/j.jenvman.2008.12.001.

Berkes, F. (2010). Devolution of environment and resources governance: Trends and future. *Environmental Conservation, 37*(4), 489–500. https://doi.org/10.1017/S037689291000072X.

Bickmore, A. K. J. (2002). *Evaluating the Co-management Institutions created by the James Bay and Northern Quebec Agreement and the Inuvialuit Final Agreement with Planning Criteria.* (Master of Urban and Regional Planning Masters), Queen's University.

Colbourne, E., Holden, J., Senciall, D., Bailey, W., Snook S., & Higdon, J. (2016). Physical oceanographic conditions on the Newfoundland and Labrador shelf during 2015. *Research document 2016/079. Canadian science advisory secretariat.* Newfoundland and Labrador Region, Fisheries and Oceans Canada.

Community Foundation of Newfoundland and Labrador. (2015). *Newfoundland and Labrador's vitalsigns.* Retrieved from http://www.mun.ca/harriscentre/vitalsigns/VitalSigns_2015-Revised.pdf

Coombs, R., Coffey, J., Dale, A., & Snook, J. (2010). *Northern shrimp policy paper: An analysis of the development and management of the nunatsiavut pandalus borealis fishery.* http://www.torngatsecretariat.ca: Torngat Secretariat.

Coombs, R., Dale, A., & Snook, J. (2011). *A socio-economic analysis of the nunatsiavut snow crab fishery.* http://www.torngatsecretariat.ca: Torngat Secretariat.

Dale, A. (2009). *Inuit Qaujimajatuqangit and adaptive co-management: A case study of narwhal co-management in Arctic Bay, Nunavut.* Retrieved June 21, 2010, from ProQuest Digital Dissertations database (AAT MR54226).

DFO. (n.d.). *Arctic char management plan for Northern Labrador.* Department of Fisheries and Ocean.

DFO. (2007). Integrated *fisheries management plan: Northern Shrimp, northeast Newfoundland, Labrador coast and Davis Strait, Department of Fisheries and Oceans, Resource Management – Atlantic, Ottawa.* http://www.dfo-mpo.gc.ca/fm-gp/peches-fisheries/ifmp-gmp/shrimp-crevette/shrimp-crevette-2007-eng.htm

DFO. (2016). *Assessment of Newfoundland and Labrador (Divisions 2HJ3KLMNOP4R) Snow Crab.* Science Advisory Report2016/013. Canadian Science Advisory Secretariat. Fisheries and Oceans Canada.

Ford, J. D., Bolton, K. C., Shirley, J., Pearce, T., Tremblay, M., & Westlake, M. (2012). Research on the human dimensions of climate change in Nunavut, Nunavik, and Nunatsiavut: A literature review and gap analysis. *Arctic, 65*(3), 289–304.

Furgal, C., Durkalec, A., Wilkes, J., Winters, K., Pilgrim, A., Webb, R., . . . , & Wilson, K. (2016). *Nunatsiavut land use, knowledge and connection to place study.* Nunatsiavut Government.

Hayes, K. (2000). *Walking together: An evaluation of renewable resource co-management in the Yukon territory.* (Master of Environmental Design Masters), University of Calgary.

Indigenous and Northern Affairs Canada. (2015). *Comprehensive claims.* Retrieved from https://www.aadnc-aandc.gc.ca/eng/1100100030577/1100100030578

Indigenous and Northern Affairs Canada. (2016). *Inuit.* Retrieved from https://www.aadnc-aandc.gc.ca/eng/1100100014187/1100100014191

Inuit Tapiriit Kanatami. (2016). *Inuit are an indigenous people living primarily in Inuit Nunangat.* Retrieved on December 12, 2016, from https://www.itk.ca/about-canadian-inuit/

IPCC. (2014). In V. R. Barros, C. B. Field, D. J. Dokken, M. D. Mastrandrea, K. J. Mach, T. E. Bilir, M. Chatterjee, K. L. Ebi, Y. O. Estrada, R. C. Genova, B. Girma, E. S. Kissel, A. N. Levy, S. MacCracken, P. R. Mastrandrea, & L. L. White (Eds.), *Climate change 2014: Impacts, adaptation, and vulnerability. Part B: Regional aspects. Contribution of working group II to the fifth assessment report of the intergovernmental panel on climate change.* Cambridge/New York: Cambridge University Press.

James Bay and Northern Quebec Agreement. (1975). Retrieved December 12, 2016, from http://www.gcc.ca/pdf/LEG000000006.pdf

Kendrick, A. (2003). Caribou co-management in northern Canada: Fostering multiple ways of knowing. In F. Berkes, J. Colding, & C. Folke (Eds.), *Navigating social-ecological systems: Building resilience for complexity and change* (pp. 241–267). Cambridge: Cambridge University Press.

Kofinas, G., Herman, S., & Meek, C. (2007). Novel problems require novel solutions: Innovation as an outcome of adaptive co-management. In D. Armitage, F. Berkes, & N. Doubleday (Eds.), *Adaptive co-management: Collaboration, learning and multi-level governance* (pp. 249–267). Vancouver: UBC Press.

Labrador Inuit Association. (1977). *Our footprints are everywhere.* Ottawa: Dollco Printing.

Labrador Inuit Land Claims Agreement. (2005). Retrieved December 10, 2016, from the Department of Indigenous and Northern Affairs Canada Web site: https://www.aadnc-aandc.gc.ca/eng/1293647179208/1293647660333

Ludwig, D. (2001). The era of management is over. *Ecosystems, 4*(8), 758–764. https://doi.org/10.1007/s10021-001-0044-x.

May, A. (1966). *Biology and fishery of Atlantic cod (Gadus morhue morhua L.) from Labrador* (PhD Thesis). Retrieved from http://collections.mun.ca/PDFs/arthurmay/May_Arthur_Phd.pdf

McDowell, G., & Ford, J. (2014). The socio-ecological dimensions of hydrocarbon development in the Disko Bay region of Greenland: Opportunities, risks, and tradeoffs. *Applied Geography, 47,* 98–110.

Michaud, W. K., Dempson, J. B., & Power, M. (2010). Changes in growth patterns of wild Arctic Charr (Salvelinus alpinus (L.)) in response to fluctuating environmental conditions. *Hydrobiologia, 650*(1), 179–191. https://doi.org/10.1007/s10750-010-0091-4.

Ministerial Advisory Panel. (2016). *Report of the ministerial advisory panel: External review of the department of fisheries and oceans' last-in, first-out (LIFO) for the Northern Shrimp fishery.* 22 June 2016. Retrieved from http://www.dfo-mpo.gc.ca/fm-gp/peches-fisheries/comm/shrimp-crevette/pdf/lifo-report-rapport-deps-eng.pdf

Nadasdy, P. (2007). Adaptive co-management and the gospel of resilience. In D. Armitage, F. Berkes, & N. Doubleday (Eds.), *Adaptive co-management: Collaboration, learning and multi-level governance* (pp. 208–227). Vancouver: UBC Press.

Natcher, D. C., & Hickey, C. G. (2002). Putting the community back into community-based resource management: A criteria and indicators approach to sustainability. *Human Organization, 61*(4), 350–363.

Natcher, D. C., Davis, S., & Hickey, C. G. (2005). Co-management: Managing relationships, not resources. *Human Organization, 64*(3), 240–250.

Norsworthy, P., Snook, J., & Whalen, J. (2012). *Scientific Research Needs: Sustainable Management of Snow Crab (Putjoti) in Labrador.* http://www.torngatsecretariat.ca: Torngat Secretariat.

Office of the Minister. (2016). *Minister LeBlanc Accepts Key Recommendations of Advisory Panel on LIO* [Statement]. Retrieved from: http://news.gc.ca/web/article-en.do?mthd=tp&crtr.page=1&nid=1095239

Pinkerton, E. (1999). Factors in overcoming barriers to implementing co-management in British Columbia salmon fisheries. *Ecology and Society, 3*(2). Retrieved December 10, 2016, from http://www.ecologyandsociety.org/vol3/iss2/art2/

Plummer, R., & Armitage, D. (2007). Crossing boundaries, crossing scales: The evolution of environment and resource co-management. *Geography Compass, 1*(4), 834–849. https://doi.org/10.1111/j.1749-8198.2007.00040.x.

Plummer, R., & Fennell, D. (2007). Exploring co-management theory: Prospects for sociobiology and reciprocal altruism. *Journal of Environmental Management, 85*(4), 944–955. https://doi.org/10.1016/j.jenvman.2006.11.003.

Rennie, H. (1989). *North Labrador and the Torngat Co-op: An exploration of Checkland's soft systems methodology through its application to fisheries development.*, Memorial University of Newfoundland. Retrieved from http://research.library.mun.ca/6644/

Rodon, T. (1998). Co-management and self-determination in Nunavut. *Polar Geography, 22*(2), 119–135. https://doi.org/10.1080/10889379809377641.

Rose, G. A. (2007). *Cod: The ecological history of the North Atlantic fisheries.* St. John's: Breakwater Books.

Schartup, A. T., Balcom, P. H., Soerensen, A. L., Gosnell, K. J., Calder, R. S. D., Mason, R. P., & Sunderland, E. M. (2015). Freshwater discharges drive high levels of methylmercury in Arctic marine biota. *Proceedings of the National Academy of Sciences of the United States of America, 112*(38), 11789–11794. https://doi.org/10.1073/pnas.1505541112.

Snook, J. (2005). *Labrador: Organized into a knot? History of the Combined Councils of Labrador. The Seventies, Eighties, Nineties and Twenty First Century.* (Local Economic Development Program Diploma), University of Waterloo, http://www.combinedcouncils.ca/home/files/pg/university_of_waterloo_thesis_paper-version_v.pdf

Snook, J. (2010). *Lessons learned from the implementation of tripartite-funded co-management boards.* (Master of Arts in Conflict Analysis and Management Thesis), Royal Roads University, Proquest. Retrieved from http://www.torngatsecretariat.ca/home/files/cat1/2010-lessons_learned_from_the_implementation_of_tripartite-funded_co-management_boards.pdf

Stevenson, M. G. (2006). The possibility of difference: Rethinking co-management. *Human Organization, 65*(2), 167–180.

Them Days. (2015). *Them Days stories of early Labrador: Nunatsiavut.* 1 Dec 2015. Happy Valley-Goose Bay: Them Days Archives and Publications, 2015.

Torngat Fish Producers Co-op. (2016). *TFPC snow crab history: 1997 to 2016.* Presentation at the 8th Annual Nunatsiavut Fisheries Workshop November, 2016. Happy Valley-Goose Bay, NL.

Torngat Joint Fisheries Board (Producer). (2011). *Legacy for life*. Retrieved from https://www.youtube.com/watch?v=Wf_cOqE3NwA

Torngat Joint Fisheries Board. (2014a). *Northern shrimp co-management in shrimp fishing area 5*. Torngat Wildlife, Plants & Fisheries Secretariat. Retrieved from: http://www. torngatsecretariat.ca/home/files/cat4/2014-northern_shrimp_recommendations_sfa_5.pdf

Torngat Joint Fisheries Board. (2014b). *Snow crab c-management in and adjacent to the Labrador Inuit settlement area*. Torngat Wildlife, Plants & Fisheries Secretariat. Retrieved from: http://www.torngatsecretariat.ca/home/files/cat5/2014_snow_crab_recommendations.pdf

Truth and Reconciliation Commission of Canada. (2015). *Honouring the truth, reconciling for the future. Summary of the final report of the truth and reconciliation commission of Canada.*

United Nations. (2008). *United Nations Declaration on the Rights of Indigenous Peoples.* http://www.un.org/esa/socdev/unpfii/documents/DRIPS_en.pdf: United Nations.

White, G. (2008). "Not the Almighty": Evaluating aboriginal influence in northern land-claim boards. *Arctic, 61*(Suppl. 1), 71–85. Retrieved November 25, 2016, from http://pubs.aina.ucalgary.ca/arctic/Arctic61-S-71.pdf

White, G., & Alcantara, C. (2016). *Institutional design and Inuit governance: Nunatsiavut and Nunavut compared*. Paper presented at the Inuit Studies Conference, St. John's, Newfoundland and Labrador.

Williamson, T. (1997). *From Sina to Sikuialuk: Our footprint. Mapping Inuit environmental knowledge in the Nain district of northern Labrador.*

Woollett, J. (2007). Labrador Inuit subsistence in the context of environmental change: An initial landscape history perspective. *American Anthropologist, 109*(1), 69–84. https://doi.org/10.1525/aa.2007.109.1.69.

Scenario Analysis for Arctic Marine Resource Policy

Niels Vestergaard

Abstract Future changes in Arctic marine ecosystems will depend as much on global climate change as on our ability to regulate and manage the exploitation pressure at sustainable levels. There is a lack of integrated, cross-sectoral ecosystem-based analysis of the Arctic marine management. The analysis would ideally include both the choices for implementing regulatory tools and how they will affect the many ecosystem-dependent values derived from them. The ability to maximize these values depends critically on the ways in which the dynamic bio-economic properties of the resources are impacted by the human behavior induced by the regulations (or lack thereof).

In this paper it is speculated about likely changes in the future Arctic fisheries based on a scenario building approach. The underlying changes to ecosystems are the climate changes which is also one of the drivers and the likely impacts in the Arctic. Other drivers can be identified but by selecting two main drivers it is possible to map four scenarios to be further analyzed. The drivers are the sectoral development of important marine sectors (fishing, shipping, mining etc.) and governance structure development. The development in each of these driving force's dimensions is uncertain and central in the analysis are risk and uncertainty. The results indicate that the future climate changes might involve relatively large changes in the ecosystem and hence fish stocks, but also that the economic outcome of fisheries depends critically upon our ability to adjust the regulatory regime to capture the values of the ecosystem services.

Keywords Scenario analysis • Arctic • Fisheries • Marine ecosystems • Uncertainty

N. Vestergaard (✉)
Department of Sociology, Environmental and Business Economics,
University of Southern Denmark, Niels Bohrs Vej 9, 6700, Esbjerg, Denmark
e-mail: nv@sam.sdu.dk

© Springer International Publishing AG 2018
N. Vestergaard et al. (eds.), *Arctic Marine Resource Governance and Development*,
Springer Polar Sciences, https://doi.org/10.1007/978-3-319-67365-3_5

Introduction

It is expected that climate change under the business as usual scenario will lead to a warming in year 2100 between two-to-six degrees Celsius in the Arctic region (IPCC 2014). In the Arctic region, humans and animals had lived and survived, mostly against all odds, during hundreds of years by integrating with and living off the Arctic marine ecosystems. History records show that changes in temperature due to natural changes have had significant impact on the living conditions and distribution of marine species in the Arctic. A recent example is the cod stock in the Davis Strait that declined so much in the early 1970s that the commercial fishery vanished in the next decades (Horsted 2000). Combining these two observations – potential large effects of climate change and 'sensitive' ecosystems – it is fair to assume that climate change in itself will make the Arctic marine ecosystems unstable with unknown future states. The opening of the Arctic marine areas with less ice coverage might lead to an increase in the exploitation level as well as the exploitation range of the marine environment and resources. This development will depend on the technological development and the price-cost relationship, where most likely – due to less ice coverage – the cost of using the Arctic marine resources will decline. The effects of climate change and the induced expected changes in both the biology and the exploitation pattern of an Arctic marine ecosystem points to the need for a sound adaptive Arctic marine policy. These ecosystems are for the next many decades in transition – as long as the temperature will continue to increase and probably longer – until stability in the ecosystems is found. During the transition period, new fishing opportunities will open up, while others will close down, because of climate-driven changes in the marine environment. The changes in fishing opportunities are due to what has been called range shift in fish stocks (Wassman et al. 2011) or 'polar shift' (McBride et al. 2014), indicating the direction of changes in range.

The current management systems which are defined and operating under assumptions of stable conditions will be challenged by continued changes in fish stock distribution and movement of fish stock towards the North and Arctic waters. The so-called mackerel war between Faroe Islands, Norway, Iceland and the EU in 2003 is an example of how current agreements do not accommodate environmental changes (Vatsov 2017).

From a scientific point of view there is a need to develop models and methodologies that can study ecosystems in transitions. The theory of bio-economic modeling is very well developed to study steady-state equilibrium including the trajectories describing the fastest way to get to the equilibrium (Clark 1980). However, development of new methods to study temporary and transient periods will be important to guide managers and politicians about the continuous changes in fishing opportunities.

The chapter will begin by setting the scene for the management issues of the Arctic marine resources followed by predictions of the future fishing options. Scenario analysis is briefly introduced and applied to the Arctic marine ecosystems. The chapter ends with two sections on future work and the conclusions.

Arctic Marine Management Issues

Future changes in Arctic marine ecosystems will depend on global climate changes and on our ability to regulate and manage the resulting exploitation pressure at sustainable levels. This will most likely be a very big challenge due to at least three main issues. The first issue is our relatively little understanding of Arctic marine ecosystems, because most of the marine area has been covered by ice – fully or partially. The result has been limited systematic data collection which can also be related to the general low commercial use of the marine resources. Our experience of exploitation is in other words 'virgin'.

The second issue is that the impact and scale of climate change on the ecosystem, i.e. how the ecosystem will change during the period, is uncertain. Temperatures have so far increased about twice as fast in the Arctic as in the mid-latitudes, a phenomenon known as "Polar amplification". As an example, predictions for year 2100 indicate that the temperature at the equator will be 1–2° C higher, while at the North Pole the increase in temperature is predicted to be 6–8° C higher. So, the uncertainty of the effects of climate change in the Arctic is, all things equal, larger than in other parts of the world.

The third issue is that there is a lack of integrated, cross-sectoral ecosystem-based analysis of the Arctic marine management. There are many Arctic marine sectors, e.g. fisheries, mining, tourism, shipping, and these will most likely grow in the future. There are many externalities across the sectors; an example is the externality related to spatial use (Kaiser et al. 2016). The changes in the management system can end up being partial, based on a poor knowledge base, relatively static in nature and not sustainable. There is a need for research and political will to change this.[1] Furthermore, an ecosystem-based analysis could include both the choices for implementing regulatory tools and how they will affect the many ecosystem-dependent values derived from them using e.g. the concept of Total Economic Value. The ability to maximize and balance these values depends critically on the ways in which the dynamic bio-economic properties of the resources are impacted by the human behavior induced by the regulations (or lack thereof).

Fishery Predictions

The changes in potential fish production are shown to most strongly mirror changes in phytoplankton production (Wassman et al. 2011). Due to both higher temperature and higher primary production (due to less ice cover and hence more sunlight) the

---

[1]The current blue growth initiative by the EU has as a central policy recommendation to integrate and coordinate the marine and maritime policy of different sectors. The same policy approach is needed for the Arctic marine ecosystems. How this is done in practice is another question.

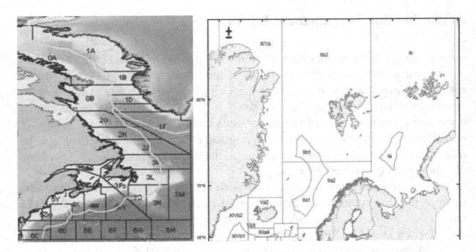

Fig. 1 NAFO Areas 0, 1 and 2 and ICES Areas I, II, V and XIV

(Sub)-Arctic fish stocks are expected to move north and some (if not most) of them to increase in stock biomass size. This will in turn form the basis for increased fishing in the region in the next many decades.

The catches in the Northeast Atlantic in ICES areas I, II, V and XIV, see Fig. 1, have the last 50 years been fluctuating between 3 and 6 mill tons. The main species are herring, capelin and cod, and it is also the catches of those species that have been fluctuating the most. In the 1970s the catches of herring and cod were low and the catches of capelin very high. In fact, the catches of herring, mackerel and cod have been increasing the last two decades and remain on a high level.

The fisheries in the Northwest Atlantic – NAFO areas 0, 1 and 2 – were in the 1960s and up to the mid 1970s dominated by cod with catch levels around 500,000 tons. From the mid 1970s and up to 1990 the catch level was around 100,000 tons and from 1990 the catch levels of cod have been very low. However, since the 1980s a shrimp fishery has been developed with catches up to 200,000 tons since early 2000s.

The increase in catches in the recent decade in the Northeast Atlantic is most likely a combination of environmental factors, e.g. a warmer ocean, and good management, while the decline in the cod fishery in the Northwest Atlantic was due to negative environmental factors, e.g. natural cooling of Davis Strait making recruitment impossible and poor management. The lesson to learn is that it is important to distinguish between long term changes (such as climate changes due to CO_2) and temporal natural changes.

The evidence in form of changes in fish distribution, spatial extension and hence in stock size and catches is still fragmented and to some extent case by case specific (IPCC 2014 and Wassmann et al. 2011). However, there are several computer based simulation models where the objective is to predict the likely long run changes in stock and fishery range and distribution resulting from climate changes.

Cheung et al. (2010) simulate future changes in maximum catch potential from global oceans by 2055 under various climate change scenarios and conclude that 'climate change may lead to large-scale redistribution of global catch potential, with an average of 30–70% increase in high-latitude regions and a drop of up to 40% in the tropics'. The changes are driven by distributional changes mainly due to changes in temperature and by changes in the productivity of the primary production. The analysis does not include changes in fishery behavior which mean that Cheung et al. (2010) will overestimate the redistribution effects. Further, there might be other ecological effects, e.g. changes in ocean chemistry leading to ocean acidification, not included in the analysis, which are predicted to have negative impacts on fishes, invertebrates and habitats. Wassman et al. (2011) and Hollowed et al. (2013) make basically the same predictions based on syntheses of current marine ecological knowledge, namely that if more ice-free periods prevail during summer in Arctic and Subarctic seas inducing increased primary and secondary production, stock biomass may increase for some commercial fish while at the same time the distribution may change as well. Besides more sunlight and increased primary production strong gradients exist from warmer, sub-Arctic waters to colder, Arctic waters, implying a high potential for species expanding into Arctic waters as temperature increases. McBride et al. (2014) conclude that "polar shift" is ongoing with a warmer climate as the driver.

Because of the lack of systematic scientific knowledge across space and time, a precautionary management approach based on ecosystem principles has been proposed including no fishing activity in the Central Arctic Ocean until the biology and ecology of the ecosystems are understood sufficiently well to allow setting scientifically sustainable catch levels (Christiansen et al. 2014). The first step towards a fishery agreement for the Arctic Ocean was taken in 2015 among the five coastal Arctic states to use interim measures such as to only authorize commercial fishing vessels based on an international fishing agreement (Anon 2015). However, for full impact of the agreement, the non-Arctic states will also need to join.

As an example of the data uncertainty in the socio-economic part of the marine ecosystem, the study of Zeller et al. (2011) of the Arctic fisheries catches in Russia, USA, and Canada is interesting. They found that cumulative fisheries catches for the FAO Statistical Area 18 (Arctic Sea, a sub statistical area of the Arctic Ocean) for the period 1950–2006 have been officially reported as 12,700 tons (t), by Russia (former Soviet Union), while no catches have been reported by USA or Canada. This compares with the reconstructed total catches of over 950,000 t, being 770,000 t by Russia, 89,000 t by USA, and 94,000 t by Canada.

To conclude this section, the scientific knowledge of the Arctic and the bordering sub-Arctic marine ecosystems is in many ways significant, but also fragmented. The scientific knowledge is in many cases not systematic (Vilhjálmsson et al. 2005), which is probably due to the ice-coverage in the past, where the cover has functioned as a protected shield of the marine ecosystems leading to poor commercial exploitation and hence, all things equal, costly data collection. Also the socioeconomic data lacks coverage and is dominated by case-studies. Because of poor data collection, there is not complete information about all economic activities.

Fundamental Uncertainty and Scenario Analysis

Based on the former section, it is fair to conclude that we are not looking at the normal decision theory case, where there is uncertainty about the value of some parameters and variables around a given mean value. The set of potential states is unknown for central factors, as well as the related probabilities. The effect of potential actions and/or their net benefits will not be easy to quantify. The uncertainty is not only related to the effects of climate change, but also to our responses with respect to the Arctic opportunities and to the general global economic and political development. In this chapter the focus is on fishing, but the development in other maritime sectors such as mining including oil and gas, shipping and Arctic tourism, is also difficult to assess and depends on external factors (price/cost relationships) as well as on internal responses (management system). Scenario analysis is a tool to be applied when there is fundamental uncertainty. By directly addressing uncertainty rather than minimizing uncertainty, scenarios encourage robust and flexible solutions that focus on more potential outcomes. Scenario analysis can inform about the sample space; how does it look like and what ranges of net benefits are likely?

Scenario planning is a method for thinking systematically and creatively about a complex future (Carpenter et al. 2006). Scenarios are sets of plausible stories, supported with data and simulations and internally consistent, about how the future might unfold from current conditions under alternative human choices. Scenarios challenge managers' assumptions about the future in a way that a single forecast cannot. Unlike the predictive modeling approach, scenario studies acknowledge the uncertainty inherent in social–ecological systems and therefore do not try to forecast the exact state of variables. Instead, comparisons among a set of contrasting scenarios are used to understand the dynamics of complex ecological-economic systems and to define a range of possibilities and uncertainties in relevant terms. Bernstein et al. (2000, 53) argue that rather than engage in prediction, a more appropriate goal of policy-relevant social science is the "identification and connection of chains of contingencies that could shape the future". Proponents of scenario building argue that the cost of mainstream policy analysis techniques that reduce complex processes to cost-benefit analysis and mathematical models is to favor precision over accuracy.

Scenarios are useful, when the uncertainty is high and the risk associated with forecasting the wrong trajectory is great. Scenario planning is based on the premise that by exploring the most divergent plausible future conditions, managers can illuminate options and risks that would otherwise be hidden or dismissed. In scenario planning, unlike decision theory, it is not necessary to assign probabilities or values to the alternatives. While the scenario development process is significantly more complex and resource intensive than regular forecast, the benefit gained is the ability to assess the robustness of alternative strategies under plausible future conditions. Decision-makers can assess the robustness of alternative policy options by determining how each policy would play out in each of the different futures.

Scenario studies can serve at least three widely accepted functions: education and public information, scientific exploration, and decision support and strategic planning (Alcamo and Henrichs 2008; Henrichs et al. 2010). In the dynamic world, scenarios can be used to highlight the opportunities and trade-offs in national and international policy debates. To make policy makers more aware of the consequences of the ongoing climate changes, scenario studies can be used to illustrate current and future changes and impacts (Victor 2012). By including alternative states of the future into a transparent problem-solving framework (Swart et al. 2004), scenario studies may help to anticipate change in social–ecological systems characterized by high levels of uncertainty and low levels of controllability (Bennett et al. 2003; Peterson et al. 2003).

There are many examples of how the use of scenario analysis has been applied successfully. The classic example comes from the well known oil and gas company, Shell. Shell used scenario analysis successfully in the 70s to predict that an alternate outcome could accrue in which a consortium of oil-producing countries would limit the production driving oil prices upward. Shell hedged against this case which allowed it to adapt more rapidly than its competitors to price increases during the mid-1970s (Van der Heijden 1996). The two most prominent examples of applications of scenario analysis to social-ecological systems are the Millennium Ecosystem Assessment (MEA) report and the work of the Intergovernmental Panel on Climate Change (IPCC). In both cases, scenario analysis is used to produce reasonable and consistent views of the likely future and to highlight where policy action is needed.

Building Scenarios

The tool-box of scenario analysis consists of several approaches, both qualitative and quantitative. Here we concentrate on the scenario-axes technique which is also applied by the MEA (Millennium Ecosystem Assessment 2005). When building scenarios it is important to find the key factors driving changes and the development, because it is often also where the fundamental uncertainties are. The scenario-axes technique (van't Klooster and van Asselt 2006) starts by identifying the drivers of change influencing the issue of interest, and providing insight about the direction for what questions politicians and managers should be asking. Then, by considering the uncertainty of key driving forces, scenarios reveal the implications of potential future trajectories.

The scenario-axes technique combines two or more key drivers of change to give a range of possible scenarios, in most cases four scenarios. The advantage of the scenario-axes technique is that within each possible scenario, detailed bio-economic modeling can be applied to analyze the outcome, e.g. the likely development of fisheries and the potential to the Arctic economy. The overall approach can include two phases. In a first phase, scenarios may be developed in an exploratory way by the scenario-axes technique, which often will be too general to serve as the basis for decision-making. Therefore in the second phase, new approaches and quantitative analysis tool may be developed using the exploration of the first phase to focus on aspects relevant to strategy and policy development.

There are many uncertainties in the Arctic about the future development. A quick brainstorm will probably produce some of the following areas of uncertainty: Climate change; International cooperation; Oil price development; Global world trade patterns; Arctic shipping and transit fees; New Arctic states; Conflict between indigenous and commercial use; Arctic maritime enforcement; Fisheries agreements; New resource recoveries.

The MEA identifies identifies nine key drivers. The economic drivers and socio-political drivers were selected in the MEA to be the ones with the highest importance and uncertainties. For our point of analysis – the Arctic marine ecosystems – the economic drivers are, among others, oil prices, trade patterns and climate changes, while the socio-political drivers are, among others, international cooperation, fishery agreements, conflicts over spatial uses and maritime enforcement.

For each driving force two attributes are selected representing two polar directions in which the drivers can go in the future. Combining these polar directions gives four possible outcomes or scenarios. The socio-political drivers can be captured by what can be called the governance axis. The governance driving force is about the degree of cooperation between actors of the Arctic marine ecosystems and stability of rules both within the Arctic and internationally. Less cooperation and stability implies an ad hoc atmosphere with less transparency, where stakeholders tend to work separately and on a unilateral basis. On the other hand, with an atmosphere of international collaboration, where nations work together to find common solutions, there is more cooperation and hence stable development of management tools and set-up.

Economic growth and resource demand defines the axis for the economic drivers and can be seen as the size of demand for Arctic natural resources and hence the size of international trade. This driving force exposes the scenarios to a broad range of potential market developments. Higher economic growth implies higher demand from more economic players for Arctic marine living and non-living resources, including increased access for trade across the Arctic Ocean. Lower economic growth implies fewer economic players interested in less marine resources. The assumption is that the world economic growth and demand will determine to a large extent the extraction level of marine resources including shipping intensity.

Combining these two axes gives us four possible scenarios as seen in Fig. 2:

Open access rush: This scenario is characterized by a poor governance structure and relatively high economic growth and demand for resources. The Arctic can be exploited by the rest of the world which needs the resources due to the high demand. The Arctic nations are not able to find ways to cooperate to control the exploitation and to secure a sustainable development of the region. Each Arctic nation defines their own marine policy which can work in some cases for some non-mobile marine resources, but it will in general give sub-optimal solutions.

Fig. 2 Possible scenarios

Underdeveloped Arctic: This scenario combines a poor governance structure with a low economic growth and demand. In this case, the Arctic does not have much interest from the rest of the world (in economic sense). Because of weak governance structure and the low global demand the Arctic will not experience a significant economic growth. One could speculate about whether this scenario is capturing the past decades of economic development in the region.

Integrated Arctic: The Arctic is integrated into the world economy with conservation of ecosystems with a stable and strong governance structure and a high economic growth. In this scenario, the Arctic development is sustainable and because of the high economic growth and demand for Arctic marine resources the Arctic economy is growing.

Arctic as a Polar Park: In this scenario, the Arctic nations are cooperating following a sustainable development path keeping the ecosystems stable and resilience. However, the economy development is lacking behind and hence relatively low. The conservation of ecosystems and relative low economic activity might develop the Arctic region into being a 'Polar park'.

This simple example of scenario building using the scenario-axes technique suggests that in a situation with high world demand and economic growth, significant changes in policies and practices towards strong governance structure, i.e. cooperation, can mitigate some but probably not all of the negative consequences of growing pressures on the Arctic marine ecosystems. These changes might be substantial and need further development and attention. If this policy challenge is not addressed, '*Open access rush*' might prevail with unstainable use of the Arctic marine ecosystem as the result.

In scenario planning, unlike decision theory, probabilities to the alternatives are not assigned. This is, however, also a weakness of scenarios analysis; it is not possible to determine which outcome and scenario that will prevail.

Discussions

Bio-economic modeling of fisheries can be applied to each of the four scenarios. Each scenario together with climate change predictions can form a basis for framing the management issue. The northward movements of fish stocks, due to climate change, create temporal and spatial externalities: How does the cooperative optimal fishery policy of shared stocks look with exogenously given fish stock movements across jurisdictions? And how will the independent fishery policy of each state look? These policy questions will have different answers, and different modeling approaches in each of the scenarios need to be applied. This is phase two of the overall exercise using the scenario exploration of the first phase to focus on aspects relevant for strategy and policy development.

The point is that scenario analysis can handle some of the fundamental uncertainties in the first stage analysis and then in the second stage quantitative modeling can be applied within each of the outcomes/scenarios to determine appropriate policy recommendations. However, the third stage is to compare the policy recommendations across scenarios to find potential robust policies, i.e. policies that are 'optimal' in several scenarios. Or if such policies don't exist, to find policies that enable reasonable responses across scenarios. Maybe the policy needs to be multidimensional to be able to accommodate as many situations as possible.

Conclusion

Scenario analysis can assist in formulating robust policies that will work to a reasonable extent in most of the scenarios. By managing the risk, scenarios that look at future paths for the Arctic marine ecosystems may help politicians to develop concrete contingency plans and strategies. Describing how and why possible futures might occur enables politicians to reflect on how political, social, and economic changes affect the society and to plan accordingly.

In this chapter the scenario-axes technique was applied to look into the future of the Arctic marine resources under climate changes and fundamental uncertainty. The tool-box of scenario analysis is big and because there is no rigorous definition and standard approaches, there exist many applications using various tools. One area not included in this chapter is participatory approaches, where the involved stakeholders are actively defining the scenarios and driving forces. This approach may give new insights into the issue and extend our understanding and knowledge and creativity towards finding stable and sustainable policies and solutions for the Arctic marine ecosystems.

References

Alcamo, J., & Henrichs, T. (2008). Chapter two towards guidelines for environmental scenario analysis. *Developments in Integrated Environmental Assessment, 2*, 13–35.

Anon. (2015). Declaration concerning the prevention of unregulated high seas fishing in the central Arctic Ocean, Oslo.

Bernstein, S., Lebow, R. N., Stein, J. G., & Weber, S. (2000). God gave physics the easy problems: Adapting social science to an unpredictable world. *European Journal of International Relations, 6*(1), 43–76.

Carpenter, S., Bennett, E., & Peterson, G. (2006). Scenarios for ecosystem services: an overview. *Ecology and Society*, 11(1).

Cheung, W. W., Lam, V. W., Sarmiento, J. L., Kearney, K., Watson, R. E. G., Zeller, D., & Pauly, D. (2010). Large-scale redistribution of maximum fisheries catch potential in the global ocean under climate change. *Global Change Biology, 16*(1), 24–35.

Christiansen, J. S., Mecklenburg, C. W., & Karamushko, O. V. (2014). Arctic marine fishes and their fisheries in light of global change. *Global Change Biology, 20*(2), 352–359.

Clark, C. W. (1990). *Mathematical bioeconomics: The optimal management of renewable resources* (2nd ed.). New York: J. Wiley.

Henrichs, T., Zurek, M., Eickhout, B., Kok, K., Raudsepp-Hearne, C., Ribeiro, T., van Vuuren, D., & Volkery, A. (2010). Scenario development and analysis for forward-looking ecosystem assessments. In *Ecosystems and human well-being: A manual for assessment practitioners* (pp. 151–219). Washington, DC: Island Press.

Hollowed, A. B., Barange, M., Beamish, R. J., Brander, K., Cochrane, K., Drinkwater, K., Foreman, M. G. G., Hare, J. A., Holt, J., Ito, S.-i., Kim, S., King, J. R., Loeng, H., MacKenzie, B. R., Mueter, F. J., Okey, T. A., Peck, M. A., Radchenko, V. I., Rice, J. C., Schirripa, M. J., Yatsu, A., & Yamanaka, Y. (2013). Projected impacts of climate change on marine fish and fisheries. *ICES Journal of Marine Science, 70*(5), 1023–1037.

Horsted, S. A. (2000). A review of the cod fisheries at Greenland, 1910–1995. *Journal of Northwest Atlantic Fishery Science, 28*, 1–109.

Intergovernmental Panel on Climate Change. (2014). *Climate change 2014: Synthesis report.* Geneva: Intergovernmental Panel on Climate Change.

Kaiser, B. A., Fernandez, L. M., & Vestergaard, N. (2016). The future of the marine Arctic: Environmental and resource economic development issues. *The Polar Journal, 6*(1), 152–168.

McBride, M. M., Dalpadado, P., Drinkwater, K. F., Godø, O. R., Hobday, A. J., Hollowed, A. B., Kristiansen, T., Subbey, S., Loeng, H., Murphy, E. J., Ressler, P. H., & Hofmann, E. E. (2014a). Krill, climate, and contrasting future scenarios for Arctic and Antarctic fisheries. *ICES Journal of Marine Science, 71*(7), 1934–1955.

Millennium Ecosystem Assessment. (2005). *Ecosystems and human well-being: Synthesis.* Washington, DC: Island Press.

Peterson, G. D., Cumming, G. S., & Carpenter, S. R. (2003). Scenario planning: A tool for conservation in an uncertain world. *Conservation Biology, 17*(2), 358–366.

Swart, R. J., Raskin, P., & Robinson, J. (2004). The problem of the future: Sustainability science and scenario analysis. *Global Environmental Change, 14*(2), 137–146.

Van der Heijden, K. (1996). *Scenarios: The art of strategic conversation.* John Wiley and Sons.

Van Notten, P. (2006). Scenario development: A typology of approaches. In *Think Scenario, Rethink Education* (pp. 69–84). Paris: OECD Publishing.

van't Klooster, S. A., & van Asselt, M. B. (2006). Practising the scenario-axes technique. *Futures, 38*(1), 15–30.

Vatsov, M., (2017). The EU's failed attempt to innovate with Regulation 1026/2012. *Marine Policy 84*, 300–305.

Victor, P. A. (2012). Growth, degrowth and climate change: A scenario analysis. *Ecological Economics, 84*, 206–212.

Vilhjálmsson, H., Hoel, A. H., Agnarsson, S., Arnason, R, Carscadden, J. E., Eide, A., Fluharty, D., Hønneland, G., Hvingel, C., Jakobsson, J., Lilly, G., Nakken, O., Radchenko, V., Ramstad, S., Schrank, W., Vestergaard, N.,Wilderbuer, T. (2005). Fisheries and aquaculture. *Arctic climate impact assessment* (pp. 691–780).

Wassmann, P., Duarte, C. M., Agusti, S., & Sejr, M. K. (2011). Footprints of climate change in the Arctic marine ecosystem. *Global Change Biology, 17*(2), 1235–1249.

Zeller, D., Booth, S., Pakhomov, E., Swartz, W., & Pauly, D. (2011). Arctic fisheries catches in Russia, USA, and Canada: Baselines for neglected ecosystems. *Polar Biology, 34*(7), 955–973.

Part III
Technology and Development

Long Run Transitions in Resource-Based Inuit Communities

Brooks A. Kaiser and Alexej Parchomenko

Abstract We discuss a multi-trophic model of socio-ecological systems. The model helps identify historical governance gaps that gained importance with the introduction of non-Inuit trading that have created lingering legacies today. In the model, humans harvest and manage a base of living natural resources. Some of the humans can organize activities that increase the resource base and/or its harvestability. These increases create capital returns, with effectiveness dependent upon governance.

A change in the terms of trade within the existing socio-ecological systems through increased global contact changed relative values. Trade induced both direct changes, e.g. in the population and in the resource base, and indirect changes through institutional gaps. Examples include Arctic fox and Bowhead whales. Early on, Inuit and outside traders saw trades as mutually beneficial. Trade also introduced new technologies (e.g. guns, traps) that lowered costs, but increased resource pressures. These transformations changed the governance needs of the socio-ecological system. New economic challenges needed changes in stewardship and institutions. In particular, institutional solutions needed to address dynamic impacts from overharvesting and to assure that trade not only increased present day but also intertemporal well-being. Stewardship over new technologies and pressures evolved too slowly compared to the rate of economic change.

Keywords Natural resource dynamics • Institutional change • Socio-ecological governance • Arctic Inuit economic development • Traditional ecological knowledge

B.A. Kaiser (✉)
Department of Sociology, Environmental and Business Economics, University of Southern Denmark, Niels Bohrs Vej 9-10, 6700, Esbjerg, Denmark

Affiliate Graduate Faculty, Department of Economics and UHERO, University of Hawaii, Manoa, HI, USA
e-mail: baka@sam.sdu.dk

A. Parchomenko
Sustainable Materials Management, VITO NV, Mol, Belgium

© Springer International Publishing AG 2018   89
N. Vestergaard et al. (eds.), *Arctic Marine Resource Governance and Development*,
Springer Polar Sciences, https://doi.org/10.1007/978-3-319-67365-3_6

Introduction

Using both archeological and historical evidence from natural resource use by Arctic Inuit communities, we discuss the role resource pressure, and lack thereof, has played in the historic development of current Inuit communities' resource governance. In particular, we focus on the roles of trade and technology in transforming well-functioning, holistic, socio-ecological systems into hybrid market and non-market economies. Understanding the economics of this transition allows one to highlight potential long run institutional gaps in governance of ecosystem services that may assist in generating unwelcome development paths. Through the combined lens of resource economics and new institutional economics, we discuss an analytical framework that illuminates several important aspects of broadly applicable principles affecting the nature and causes of growth and institutional change.

Simple historical economies[1] provide opportunities for insights into economic theory that are more difficult to disentangle in more complex globally integrated economies (e.g. Brander and Taylor 1998; Fisk and Shand 1970; Kaiser and Roumasset 2014; Taylor 2011). Such simple historical cases, however, often lack sufficiently delineable evidence covering the range of evolutionary pressures needed to inform broadly across a developmental spectrum due to the lack of written historical records. For example, in many models of simple economies, resource development is inextricably and monotonically linked to population dynamics (e.g. Brander and Taylor 1998). This is potentially misleading in that these models do not allow much scope for shifts in such important dynamic factors in resource use as trade opportunities, technological progress, or development of a capital-intensive class. The relatively well-documented and dramatic transition from resource-based, closed, Inuit economies and societies through their economic and institutional integration into the high-GDP economies of the US, Canada and Denmark (Greenland) provides opportunity to investigate the dynamic relationships between a resource base and the population generating *total economic value*[2] from it, including governance and institutional structures.

We take Kaiser and Roumasset's (2014) model as our starting point, combining theories of costs and benefits of governance and institutional change (New Institutional Economics, see e.g. (Ménard and Shirley 2005)) and ecological models of resource dependency (Resource Economics, see e.g. Costanza et al. 2014; Costanza et al. 1993; Van den Bergh 2002). In this dynamic model of a resource-based

---

[1]By simple economies, we refer to economic conditions where the number of goods and services produced and/or traded are significantly limited by opportunity and availability of resources and technology.

[2]*Total economic value* is defined as the value of the net benefits to society, fully measured. This includes direct and indirect use values as well as non-use values. It is particularly appropriate for use in cases such as this Inuit economic analysis where market activities are highly incomplete measures of value. See Emerton 2016 for a more detailed overview.

economy, governance co-evolves with production, specialization, and population. They apply the model to the case of Hawaiian economic development, with a focus on the gains from specialization and the development of a productive managerial elite. The introduction of trade at the end of the eighteenth century shifts returns to capital and trade, so that the institutional evolution transforms from increasing efforts to capture hierarchical benefits of control to efforts to increase decentralized benefits from information. The Inuit and Hawaiian cases have many similarities – they are both historically isolated communities dependent on a limited supply of natural resources. Distinctions between the two communities, however, especially with respect to the building, use and management of capital, enable comparative analysis of the roles of trade and technology in socio-ecological systems under stress from dramatic shifts in values and/or productivity. By bringing anthropological and scientific research to bear in our economic model (e.g. McGhee 2007; Raghavan et al. 2014), we separate and examine extensive growth, intensive growth, and trade and technological evolution as it depends on an initial resource base.

We model an economic system in which a composite resource stock, or ecosystem, is harvested for use by a human population, where the resource stock is subject to natural biophysical limits. These biophysical constraints have dynamic feedback effects on the stock. These constraints are affected by the (transactions) costs of managing and governing harvest from a composite ecosystem. The ecosystem stock is harvested for three purposes: consumption by the (endogenous) laborer (subsistence-motivated) portion of the population, export in exchange for goods external to the resource base, and/or capital (wealth) accumulation that sustains a capital-endowed, governing class whose contribution to growth stems primarily from increased returns to capital. The governing class is considered, in the language of ecology, a 'top predator'. They are the source of capital accumulation and technological change. These multi-trophic interconnections differentiate and broaden the story from primarily open-economy discussions where resources are providing different returns from physical and/or human capital (Carboni and Russu 2013; Eliasson and Turnovsky 2004; Lopez et al. 2007).

In Inuit Arctic communities, this top trophic level takes the form of Human Capital. This human capital manifests itself as Traditional Ecological Knowledge or TEK. TEK can play two important roles. It can directly increase the efficacy of resource harvesting (e.g. improved harpooning), thus increasing labor productivity (catchability), increasing pressure on the resource. It can otherwise increase the base resource's productivity (e.g. improving use of marine mammal parts for sustenance and survival – capital deepening), potentially increasing or decreasing resource pressures, depending on how this shifts relative growth rates of the resource and the population – that is, whether life (wealth) is improving in such a way that population expansion or contraction follows.[3] This latter use of capital may allow for reduction in intraspecific competition of the human population (e.g. through

---

[3]Important exogenous population shifts from contact with new diseases can also be considered in this framework.

territorial expansion into unused resources). Changes in (exogenous) resource values, harvesting and governance costs for common property resources, and costs of enforcing resource use for wealth accumulation (e.g. protecting an elite or governing class) and/or trade (e.g. regulation of markets or other governance of exchange) are investigated in order to explore the co-evolution of governance structures and resource pressures. In our exploration, we present sparse evidence substantiating stylized facts about past and recent historical Inuit economic development.

Dynamic Theory of Resource Use and Institutional Change

We begin with a simplified exposition of the Inuit case in the context of extensive growth, intensive growth, technology and trade in resource use for a resource-based economy. To focus on the most resource-constrained communities, we limit our analysis to those communities existing above the tree line. The isolation of Arctic Inuit communities to groups facing similar resource limitations and technological challenges and the severity and seasonal extremes of the climate meant that there were severe limitations in providing new technologies or trade opportunities for much of the communities' development. The substantial levels of internal trade and connectivity amongst historical Inuit communities (Aporta 2009; M. W. Betts 2007) allow us to simplify the modeling to that of a pan-Arctic Inuit community, where the variations in communities are explored within the model framework to support the model findings. This allows for clearer insights into the intertwined relationship between institutional change and stages of growth.

Archaeological and Historical Record in the Context of Development

The Thule, unlike their predecessors, appear to have successfully harnessed dogs for transportation, creating Pan-Arctic capabilities for (seasonal) communication, trade and exchange throughout the Inuit Arctic (Cooper et al. 2016; Morey and Aaris-Sørensen 2002) beginning around the thirteenth century AD.[4] As this trade required highly stable ice coverage of the land and seascapes, it remained limited to populations with similar resource endowments and did not greatly expand the

---

[4]In the Inuit Arctic as a whole, there have been two main population waves, with the second, Thule Inuit population replacing an original Arctic culture in about the thirteenth century CE (Raghavan et al. 2014). As there is archaeological evidence of temporal population overlap but no virtually no genetic mixing (e.g. Moltke et al. 2015), it is understood that the transition from the early to the current cultures is a full displacement of one set of institutions and technologies by another. With archaeological efforts still rather preliminary regarding this dramatic transition, we concentrate here on the second, integrated population of Thule Inuit, hereafter referred to as Inuit.

diversity of ecosystem services available for use. The descendant Inuit populations include several small communities from Eastern Siberia through Greenland, with some distinctions in resource abundancies and accompanying technologies. We focus here on the Inuit population living above the tree line and on the earth covered in tundra (and sea covered at least seasonally in ice). For these individuals, most of their combined sustenance (food, clothing, shelter) came from the sea (marine or aerial).[5] Coltrain (2009) calculates possible diet compositions from a number of Thule Inuit remains that suggest their ancient diets consisted of 81–84% marine foods.

While some food and resource material could be harvested individually, the lack of natural capital suitable for generating clothing, shelter, and heat/light in forms that could be individually collected meant that coordinated activities to acquire, in particular, large marine mammals (whales and walrus specifically) would be greatly valuable from the onset of any community. Ryan (2011) highlights the support that Coltrain et al. (2004) and Coltrain (2009) give to other research through their bioarchaeological findings that link Inuit status differences to bowhead whale distributions. These bioarchaeological findings show variability of the distribution of bowhead whales in ancient Inuit diets (Coltrain 2009; Coltrain et al. 2004).

Similarly, we should consider that the broad biophysical carrying capacity of the environment, even when carefully used for sustained human population's needs, was low overall, with seasonal and/or specific abundances. Before significant trade with non-Inuit communities, the general lack of such resources as forests, metals, energy sources, and beasts of burden necessarily limited the need and development of broad technological progress; the major and important technological innovations are considered the dogsled and whaling tools including the harpoon, kayak (individual skin boat), and umiak (large skin boat) (J. Anderson 2011).

Thule Migration and Trade

The uniformity of the language of North American and Greenlandic Inuit indicates that eastward expansion is recent (in the past 1000 years) and that such high degree of language similarity is the result of dispersion from a single source (McGhee 2007). Genetic analysis further supports this (Raghavan et al. 2014). Hypotheses for this dispersion can be categorized in two main camps encompassed in our model: population growth that required increased resources and perhaps followed marine mammal migrations (extensive resource-limited growth), or trade-driven enterprise, primarily in search of more metals, to increase harvestability

---

[5]The cases of Arctic populations utilizing land based resources such as reindeer and caribou or forests are also interesting, particularly in that wealth storage "on the hoof" was feasible and thus wealth accumulation and social stratification were more prominent components of population growth and development. We save these cases for future research.

of sufficiently abundant resources (and possibly result in population growth). We suggest the latter was important for Pan-Arctic expansion, while the former may have prevented southern migration, but that TEK may have been equally important or even more so.

Push factors for Arctic migration include inter-community hostility[6] from the south (i.e. by the Dene Indians, who occupied the forested areas) that then favored eastward over southern migration (McGhee 2007), as did the technology of dog-sled transport. We therefore expect that technological transferability channeled migration between today's Alaska on to Hudson's Bay and across the Davis Strait to Greenland, exploiting the relatively rich resource base of the Arctic marine waters with developed TEK skills, in spite of the lack of land-based resources.[7]

Migration also may have been tied more directly to the resource base. Whale migration is a natural cycle that extends the regional abundance of resources to areas such as western Canadian Arctic. Hunting just one whale can provide up to 50 tons of meat and blubber and explains the establishment of more densely populated areas and villages of up to 100 people at more northern coasts (McGhee 2007). Thus, replication of existing communities (extensive growth) could develop along a long shoreline with access to migrating whales (Higdon 2008).

Apart from seasonal oscillation of whale abundances, long-term shifts in abundance may be another reason for some migration.[8] Environmental shifts, e.g. during the small ice age in the 17th and 18th centuries and before, could have caused earlier Inuit to migrate to follow the changing patterns of whales (Wenzel 2009). Limiting factors for expansion were overcome with regional trade (Rasic 2016) and with innovations such as tailored skin, bow and arrow, etc.; this technology was crucial for survival in the harsh Arctic climate. This supported a hunting way of life for both the earlier people who came to North America from the east (McGhee 2007) and the modern Inuit.

Not only do geography, currents and animal migration patterns generate different productivity levels across these similar environments, they also present different accessibilities for early trade. Trade advanced into the Inuit communities both from the east (Higdon 2008) and west (Bockstoce 2009), so that the last arena for direct trade with more developed economies was in the central Canadian archipelago (McGhee 2017). Today, this area remains the least integrated into global market activity and the most dependent on its resource base for subsistence. It has not yet been significantly infiltrated even by (extractive) commercial fishing. From Jan 1, 2012 through Mar 31, 2017, only two fishing vessels over 15 m length are known

---

[6]Existence of such hostility is indicated by archaeological evidence of armor and bows (McGhee 2007).

[7]This is supported as well by the discussion in M. W. Betts and Friesen (2004) of the Thule Inuit development in the Mackenzie Delta of the Canadian Arctic, whereby existing TEK was applied in ideal environmental conditions to enable an increase in prosperity.

[8]That whales migrate and adapt to changes is shown by a recent case in Greenland. In the 1930s the Beluga whales moved northward, due to the warming of the waters of West Greenland (Freeman et al. 1998).

to have operated in any part of the archipelago.[9] The vessels are the Kiviuq 1, belonging to Nunavut's Arctic Fishery Alliance (http://www.arcticfisheryalliance.com/vessels.html), which has fished from the east as far to the northwest as Bathurst and Cornwallis Islands, and Frosti, a Vancouver-based trawler that has fished as far east as Ulukhaktok on Victoria Island. No commercial fishing has occurred since 2012 from the western side of Victoria Island to the eastern side of Bathurst Island and down the eastern side of Baffin Island. The area meanwhile is slowly becoming open to tourism aiming to benefit from its distinctions from market driven economic forces.[10]

On the other end of the trading spectrum is one of the globe's most productive marine environments: the southern coast of Alaska (McGhee 2007). Exploited for thousands of years (Finney et al. 2002), these fishing grounds were also connectable via currents and migration patterns to Asia. McGhee (2007) theorizes that around 2000 years ago, access to metal tools of Bering Sea Aborigines produced a shift induced by trade, probably motivated by demand for furs and other Arctic products by an elite class of Asian societies. Further, this allowed Inuit to form new tools by utilizing the iron blades, i.e. carving ivory (McGhee 2007). This would mean that eastward expansion already occurred in part with the possession of metal tools and improved tools. Moreover, the prospects of access to iron in the Eastern Arctic via iron ore or traded iron further support the role of iron as pull-factors for the movement to the east (McGhee 2007). Indeed, some early trade in the east is likely to have occurred, as supported by the less fanciful medieval accounts of Thule and treasures such as Greenland kayaks on display at the cathedral of Oslo in the sixteenth century, possibly dating back to the thirteenth century[11] (Vaughn 1994). Guns, however, were unknown before the early nineteenth century (Bockstoce 2009).

Technology

The separation of gains from trade, technology, and the lack of significant physical capitalization is a large part of what makes the Inuit case so useful in elucidating our model. Significant gains from TEK came through production of goods that could not be individually produced: group-organized labor increased both quantity and

---

[9]This can be seen directly via globalfishingwatch.com's map of fishing activities tracked by Automatic Identification Systems.

[10]The question of whether tourism can grow while being any less disruptive than other trade introductions remains an open one, with a growing research agenda (Zeppel 2006).

[11]Olaus Magnus writes in 1515 that he saw two Greenland kayaks hanging in the cathedral at Oslo. He was told that they had been taken by King Hakon IV of Norway (1217–63). The same king apparently gifted Edward III of England a polar bear in 1252 that was kept at the Tower of London (Vaughn 1994).

quality of subsistence. In particular, TEK enabled the transformation of the frozen Arctic tundra into self-sufficient integrated resource use communities. Gains from specialization and intensification in this system could be achieved, but there was little scope for wealth storage outside of human capital. While all historical Inuit held TEK capital that allowed for individual (family unit) survival, stratification of TEK and related physical capital was known to exist (Ryan 2011) and has been particularly related to whale harvesting capabilities (Ryan 2011; Stern 2013). We focus here on the group investment in individual TEK and related physical capital that fosters the ability to catch and use large marine mammals. While important, this is only one component of Inuit activities, and communities engaged in communal whaling to greater and lesser degrees. The limited need for collective action and limited presence of physical capital restricted the growth of any elite. Any elite would have lacked a base upon which to survive without contributing directly to productivity; the harvest levels depended directly on TEK capital as an active component of the harvest (and was increased through the practice and learning afforded by harvest activities) rather than a passive investment cost. Thus the productivity of the TEK and related capital is directly viewed as a productive part of the socio-ecological system.

Furthermore, the value of TEK was infinitely depreciable, requiring intergenerational transfer. The investment in 'toys' which mimicked harvest and survival needs illustrates this (Laugrand and Oosten 2008). Continued TEK is highly dependent on continued use of the resource base, so that trade not only supplemented TEK with enhancing technologies (e.g. guns, traps), but also may have increased the rate of its deterioration as the cost of this intergenerational transfer increased in relative terms, and shifted to incorporate the new technologies. Depreciation of TEK and introduction of new hunting technologies is therefore likely to have happened in waves. For example, resolution of conflicting accounts from Alaska now indicates that the first firearms appeared in that region during the 1820s and most likely dissipated east through native trade, while a second wave came after 1848, when the British whaling and trading fleet arrived and economic activity became more regular (Bockstoce 2009).

As resource pressures increase, Kaiser and Roumasset (2014) argue that returns to specialization, intensification, capitalization, technology and governance require additional centralization of authority and decision-making, which, in small populations with limited opportunities for external economies of trade, can be developed through hierarchy, if capital (wealth) accumulation is possible. A lack of managerial gains from formation of a governing elite, however, can limit hierarchical and/or institutional development, because limiting today's production in search of higher gains tomorrow is neither necessary nor possible.

In the Inuit Arctic case, without much storage of wealth possible, governance gains to hierarchy were limited – overharvesting was not contemplated as a human-induced problem; the gods, if properly requested, provided.[12] While religious rules governed the hunt,[13] these rules were not functional taboos or otherwise limiting in the economic sense of governing community property. They were and continue to be more clearly associated with coordinated risk-sharing and risk abatement (see e.g. Ford et al. 2006, 2008). Thus, we interpret managerial investment (TEK) and related governance in this case as referring to the intensity and extent of ecosystem extraction rather than the quantity of any one species harvested. In terms of governance needs, this approach is akin to providing insights into ecosystem management rather than single-species management in current context. That is, we consider the resource base a composite good (e.g. marine mammals, birds, fish, etc) that can, through increasing levels of (costly to acquire) ecological knowledge, be utilized at higher and higher levels. Over- or improper use of any one component of the composite may jeopardize production of the whole, but dispersion of use within the composite can reduce risks (through diversification) and waste within the system.[14] TEK is a key component of successful dispersion of ecosystem use. Further, this approach is consistent with Inuit management interests that incorporate broader aspects of the resource than population counts (Tyrrell 2007).

Illustration of the Dynamic Forces of Ecosystem Interactions, Trade and TEK

Figure 1 illustrates the intertemporal choices and outcome directions in a resource based economy. First, the (composite) resource base is either harvested or it is allowed to grow for tomorrow.

If harvested, portions of it may be consumed (feeding a subsistence population), traded, or 'invested'. Investment feeds knowledge; human capital transfer is enabled at the cost of current direct production. Knowledge can increase future harvestability (lower harvest costs or increase dispersion options) or increase the future resource base, with differing impacts on long run growth. Trade severs ecological constraints by introducing goods and services from outside the existing resource base. If portions of the resource base that are not currently in use by the society are traded, trade can be win-win, as long as dynamic ecosystem impacts do not deteriorate the

---

[12]The early Norse colonies in Greenland provide a contrast regarding attempts to support hierarchy and generate surplus elite in Arctic conditions. The Norwegian church and state taxed the colonies heavily with little in return, undoubtedly a contributing factor in their mysterious demise in the sixteenth century (Kintisch 2016; Vaughn 1994).

[13]Alaskan Inuit, e.g., would hunt a whale guided by a shaman, including a distinguished hunter and certain rules and rituals had to be followed before the hunt, so that the whale would willingly give himself to the hunters (Freeman et al. 1998).

[14]This composite ecosystem might be further broken down into its hedonic components, which would facilitate discussion of the transformation of relative resource values within the system. We save this for separate research.

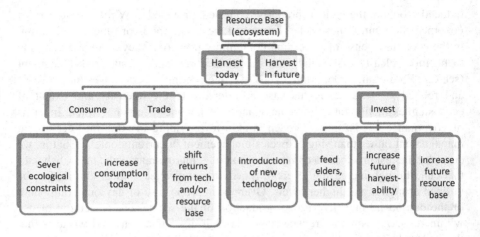

Fig. 1 Multi-trophic resource-based economy

productive capacity of the system. Otherwise, trade may increase or decrease overall well-being depending on the terms of trade. Exchange must cover the dynamic costs of replacing the resource in the socio-ecological system to be mutually beneficial. Trade that introduces technology used to increase the harvest beyond sustainable levels in the socio-ecological system – e.g. commercial whaling in the Arctic – may not be welfare enhancing.

Intertemporal Considerations of the Multi-trophic System

Value of the Resource over Time

The value of the resource to society depends upon its division. Immediate consumption generates the initial level of well-being today, and provides the base for human population growth for tomorrow. The share to trade is exported for current additional benefit. It can enhance or detract from future growth in capital or a subsistence population dependent on whether consumption or investment goods are purchased. The share of the resource base devoted to supporting TEK capital creates value through investment in the human capital (TEK) of the population. The sum total determines the remaining resource base available for growth (or replenishment) and future value.

Value from Subsistence Consumption

The marginal value of consumption of the resource for human subsistence by individual harvesters will exhibit the standard properties of demand, so that increased consumption is expected to increase wellbeing at a decreasing rate and so that diversity of consumption is valuable. Thus trade is of current interest as it can increase well-being through diversity of consumption.

In addition to the current benefits of consumption and trade, the current harvest affects the future resource base and human population. The resource base grows as a positive function of ecosystem capacity and negative function of human harvest. The human population tomorrow is determined by the ability to convert consumption to growth (via an intrinsic growth rate), the death rate of the population, and the intraspecific rate of competition, i.e. the rate at which members of the population compete for the same resources. When there are sufficient resources so that there is no deadly competition for the resource base, it must be that the population either can simply expand with extensive growth (e.g. into new resource-rich areas) or reproduces so slowly as a function of a low net fertility rate that competition does not evolve. Further, this is a function of investment decisions by the capital holders, where capital accumulation (through TEK) can counteract crowding by resource-increasing investment and an expanding production possibilities frontier.

Figure 2 provides an illustrative overview of this mid-trophic layer of the socio-ecological system. In it, dynamic self-sufficiency, TEK and technology are on display. The reproducing family unit is shown with its dogs, hunting gear, and clothing from marine mammals. This self-sufficiency is reproducible but there is little scope for wealth accumulation.

Value from Human Capital Formation

The marginal benefit (utility) of capital accumulation may be described generally as a decreasing benefit function where total benefit from TEK (capital) shifts via changes in the benefits of wealth (perhaps prestige, power, or access to luxury goods). The share of the resource base that goes to capital contributes to current well-being through a technical transformation from the resource into capital allowing for new capital (ecological knowledge) formation. The opportunity costs of TEK are the reduction in resource availability for trade or direct consumption (and direct population growth). Capital investment or exogenous shifts in technology change the rate of this transformation from resource to capital. We presume that investment and/or contact with others through trade could increase the ability to transform the resource into capital value and would thus increase the amount of capital available to the system. Capital depreciates; one may consider this depreciation rate to be the mortality rate for the holders of TEK.

Value from Trade

The marginal benefit (utility) of the resource as an export commodity should also exhibit decreasing marginal returns, which shift via changes in opportunities for trade. The current net benefits to trade must be balanced against the lack of

Esquimaux Indians of the Coast of Labrador

Fig. 2 The mid-trophic layer of the socio-ecological system, illustrated (Image credit: "Esquimaux Indians of the Coast of Labrador" communicated by a Moravian missionary, drawn by Garret, engraved by Chapman, published by C. Jones, October 17, 1818. From Charles de Volpi, *Newfoundland: A Pictorial Record* 15 (Sherbrooke, Quebec: Longman Canada Limited, ©1972))

availability of that resource for either consumption or capital purposes. The human subsistence population and capital growth will be lower with more trade, unless trade replaces the lost resource base with new opportunities or reduces the effort required for resource extraction (shifts technology). These may include direct food supplies or changes in technology that affect the catchability coefficient. Control of the resource for trade and the distribution of returns from trade are then important factors in support for the institutional structure of the economy.

Costs

The benefits of the harvest are countered by the costs of the harvest and costs of harvest governance (here, TEK-influenced dispersion), which apply regardless of end use of the resource. Here we discuss the relationships between marginal costs and the working of the resource dependent system. Note that all of these costs may also change through exogenous shocks over time.

Current Harvest Costs and Costs of Dispersion

The per-unit cost of harvest may be a function of the resource population and/or capital stock, though for simplicity we consider it exogenous. The effects of changes in the per unit enforcement costs of the shares to capital accumulation and trade (described further below) are expected to behave very similarly. The sensitivity of the effectiveness of harvest technology (essentially a 'catchability factor') to capital stock can be expected to act in the same manner as decreasing costs from increases in the stock (and vice-versa). Furthermore, the dynamics of endogenous harvest costs are explored in the renewable resources literature, so our discussion draws on these findings (see e.g. Brander and Taylor 1998; Eliasson and Turnovsky 2004; Lopez et al. 2007; Squires and Vestergaard 2013).

In general, failure to limit the harvest today is expected to result in overharvesting and inefficient allocation of the resource base across time (Hardin 1968). This will not be the case, however, if resource pressure is sufficiently low that open access does not jeopardize the future harvest (Kaiser and Roumasset 2014). The possibility of illicit harvesting is also low in the case of large marine mammal harvests, as collective action and specialized human capital are necessary. We therefore interpret governance costs of the harvest as broader than direct limitation of harvest that is generally the focus (see e.g. Clark 2005). As we consider the resource base a composite good, the application of TEK to harvesting greater amounts within the composite resource (increasing dispersion of ecosystem use) is increasingly costly.

An example of this might be as follows. A community is harvesting seals with essentially constant returns to scale and no need for limits due to abundance of seals and low human population. Arctic fox are not harvested – dispersion in ecosystem use is relatively low. Trade arises; fox fur becomes valuable for trade. TEK may be developed to also trap fox (TEK increases through increasing the share of the resource to capital) – in turn the use of the ecosystem becomes more broadly dispersed, and costs of this ecosystem harvesting as a whole are now higher.

Current Enforcement Costs of Non-consumption

We must also consider the costs of enforcing decisions regarding the shares to consumption, capital and trade. We do so separately, to allow flexibility in considering how shifts in external prices for the resource base may create different pressures and costs on enforcement, and in order to better reflect on the role of management and human capital. We assume that costs of portioning off the share to TEK holders, or capital accumulation, are non-decreasing in the share. This is because the more of the resource that is taken from direct consumption, the greater the monitoring costs and related costs of ensuring that the capital is efficiently allocated. We also assume that the costs are non-decreasing in the number of people needing to cooperate, as one expects in commons problems (Field 1989; Glaeser and Shleifer 2003; Kaiser and Roumasset 2014). Costs of enforcing a share to trade may be considered similarly non-decreasing in the share to trade, but also with respect to population levels. This is because the opportunity cost of trade over consumption will increase and/or more individuals involved in trade result in more parties to monitor who might rather consume the resource.

Resource Harvest

The resource is harvested by the population at a per capita rate that determines its continued growth through its ability to convert a unit of resource into sustenance (for example the hunting success rate of a whale population), with an additional consideration. That is, this ability is a function of TEK and related physical capital. With respect to the latter consideration, we in general expect increases in capital investment (and/or harvest technology) to increase catchability of the resource population. The more abundant the resource, the easier the harvest, ceteris paribus, as one would expect (Clark 2005; Clark and Munro 1975). The transfer of TEK to catchability is what we consider the dispersion of use of the ecosystem. This transfer is such that increasing TEK increases the usability of the composite resource – in other words, it reflects breadth of use, or how much of the composite ecosystem base can be harvested and used. If TEK fails in the extreme to enable any transformation of the ecosystem base into sustenance, for example if all hunting techniques become ineffective, then use of the ecosystem falls to nothing and there can be no harvest, as no component of the composite resource can be made valuable. If TEK is fully implementable for the environmental conditions, then the use of the ecosystem – its catchability – is at its maximum rate for the existing technology – which may or may not be sustainable. We assume that this dispersive use is increasing in TEK investment. Achieving higher levels of dispersive use is shown in the Arctic context to require coordination activities by the TEK holders, as with whaling operations.

Figure 3 illustrates an Inuit whale hunt. The image reflects the need for a steady and skilled crew, and the collaborative advantages in acquiring not only the whale but also other marine mammals.

Discussion of the Socio-ecological System Overall

In short, there is an underlying maximization problem of the present value of the ecosystem for subsistence, trade, and capital formation across an infinite time horizon. This maximization is subject to constraints on the biological reproduction of the composite ecosystem resource, the biological reproduction of the human

Fig. 3 The whale hunt (Image credit: Frédéric Back, from *Inuit: Glimpses of an Arctic Past*, by David Morrison and Geroges-Hébert Germain, Canadian Museum of Civilization, S2002-4627)

population, who are dependent on that resource, and the human (TEK) capital formation, which is also dependent on the resource as well as the population to harvest it.

Trade removes subsistence resources from the population and may effectively lower the cost of ecosystem harvest through the introduction of new technologies. At the extreme, trade may provide a partial-equilibrium[15] exit strategy from the subsistence life; if the value of the traded resources is sufficiently high that it entirely compensates for long run subsistence benefits, then a corner solution to the maximization problem is to sell off sufficient ecosystem resources to move to a market based existence outside the ecosystem. This describes the underlying fundamentals of what is happening when Inuit decide, individually or as a community, to enter fully into a market-based economy and move away from direct dependence on the resource base.

Exit by TEK holders hastens the capital depreciation as well. Thus, by couching the returns to trade in terms of the present value of foregone well-being to the community, one can examine the tradeoffs involved in making this exit at the individual and/or community level. This allows one to begin to account for the true long run tradeoffs that Inuit face today in maintaining cultural values vs. shifting to new forms of economic activity.

TEK in turn might effectively expand the carrying capacity of the ecosystem and/or the technological capabilities of harvest (its harvest intensity within one

---

[15]This is a partial equilibrium outcome because the new (market based) ultimately must be supported by resources from some other ecosystem.

aspect of the ecosystem or through the dispersion of use). This can ease the population constraints and allow for additional growth, but may also increase harvest pressures on the ecosystem and threaten a long run sustainable growth path. Thus the state variables are the resource (ecosystem) quantity, the human laborer population, and the level of TEK (as manifested through managerial elite holders of that knowledge).

A control decision thus exists for impact on each of the three trophic levels. These are:

1. the dispersion of ecosystem resource used,
2. the share of ecosystem resource used in trade, and
3. the share of ecosystem resource used for TEK and related physical capital.

The control decisions impose associated costs that are non-decreasing in the intensity of control. The combination of the choices over the controls and the impact of their costs as values for consumption, trade and capital shift determine the community path over time. In addition to system exit, a sustainable balanced growth path requires that use of the composite ecosystem resource remains slow enough to avoid collapse but intense enough to allow growth in well-being. Growth in capital (TEK) may be able to increase well-being, but only if balance can be maintained. In other words, TEK that is used to expand the resource base or the (dynamically supportable) dispersive use of the ecosystem is more likely to increase sustainable well-being than TEK that increases intensity of use of a single component of the ecosystem.

The system linkages of the resource-based system that determine whether economic conditions warrant system exit (without collapse), a sustainable, technology-dependent balanced growth system, or system collapse (or related system cycles that generate losses to dynamic welfare), can be expressed in three interrelated shadow values, one for each level of the multi-trophic system.

At the base, the shadow value of the ecosystem – that is, the true opportunity cost of using a valuable unit of the ecosystem – is equal to the full marginal costs of acquiring that unit of the resource. This full marginal cost includes harvest costs, costs from dispersive use of the system, and costs of TEK investment, net of its value from trade (exiting the system). Thus, shifts in any one of these costs or values will permeate throughout the system through changes in the value of the ecosystem base.

Similarly, the shadow value on the human population base is equal to the value of the resource for subsistence, net of its value for trade and the costs of the resource devoted to TEK capital, scaled by the intrinsic growth rate of the human population. In other words, the shadow value is the additional value from population growth foregone from devoting another unit of the resource to trade or capital.

Finally, the shadow value on TEK formation is the foregone returns from using the marginal unit for trade or human population growth.

Applied Analysis for Inuit Communities

The multi-dimensional system presented here is rich in potential detail and will be explored at greater length in additional work, but we focus here only on pieces where the Inuit case is particularly revealing to the overall discussion regarding governance issues that pertain to long run sustainable development from an ecosystem base. We turn therefore to examine these interrelated aspects of potential interest in that context:

1. shifts in relative values for subsistence, trade, and TEK capital;
2. changes in returns to cooperation (requiring management);
3. changes in returns to TEK affecting dispersive ecosystem use and/or ecosystem carrying capacity; and
4. conditions governing exit from the ecosystem.

We examine the interrelated effects of these aspects on the limited historical development of governance by the Inuit through two cases: first, the introduction of trade via Russian and other European fur interests, and second, whales and whaling.

The Fur Trade

Trade between Russia and subarctic portions of Alaska (mainly the Aleutian Islands) for furs was underway by the seventeenth century (Bockstoce 2009). The sale of Alaska to the Americans in 1867 expanded this trading network, already well developed in Southern Alaska and the Aleutians by Russian, American and British traders, and expanding north since reports of Captain Cook's foray through the Bering Strait sparked interest in marine and land animal triangular trade with Asia (Bockstoce 2009). The map in Fig. 4 illustrates the lack of expansion and connectivity even at the end of the eighteenth century; Cook's map was certainly enhanced from sharing by Russian traders (Stern 2016) but still just scratches at the doorway of the Arctic tundra.

The Hudson Bay Company was founded for the exploitation of the Canadian fur trade in 1670. Yet trade was so controlled by the monopoly at the frontier that as late as the start of the second world war there were still communities that had only ever interacted with one or two traders (Poncins 1941). Danish colonization of Greenland dates from 1721, with earlier contact by fishers and traders. Danish control of Greenlandic trade outside of Inuit communities was complete enough to consider Greenland a closed economy (Nuttall 2005). Trade for furs focused on a few goods. These goods included metal goods including kettles and knives, traps, saws, pans, guns and ammunition; foodstuffs including flour, sugar, and molasses; and tobacco, matches, and sewing implements (Bockstoce 2009). Generally contraband alcohol was also prized for trade, though access was limited. Even still, the consequences of these few items infiltrating Inuit communities over time were dramatic. Changes

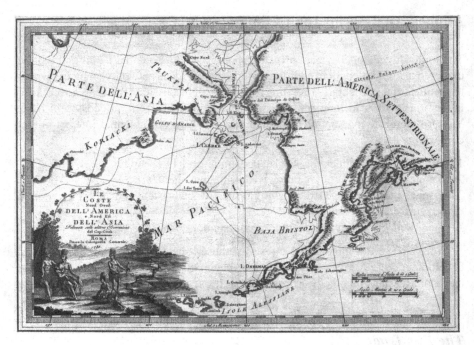

Fig. 4 Contemporary map illustrating Captain James Cook's voyage in the Bering Strait (Image credit: Cassini, Giovanni Maria, *Nuovo Atlante Geografico Universale Delineato Sulle Ultime Osservazioni*, V. 3, Rome: Calcogrfia camerale, 1798, (public domain) provided to Wikipedia Commons by Geographicus Rare Antique Maps)

in relative values and changes in returns to individual harvest efforts interacted with changing impacts from increased dispersive use of the ecosystem to shift institutional needs and individual vs. group incentives.

With the change in opportunities for trade that accompanied increased external contact, the benefits of coordination had relatively less ability to increase individual welfare. The switch in values to tradable goods that were individually harvestable, in particular Arctic fox, effectively reduced the need for coordinated hunting benefits for marine mammals and led to institutional changes favoring decentralization of decision-making, while increasing centralization of authority in the governance of property rights and to dramatic reductions in TEK. Trade in technology, particularly for harpoon improvements, guns, and fox traps, reduced effort costs of both group and individual harvests.

Inuit had relatively little direct use for Arctic fox prior to trade with the external world; generally they were not heavily harvested for subsistence use (Bailey 1993; Damas 2002); there is little evidence of investment in TEK or physical capital (traps) for harvesting foxes prior to external contact. The fox has also now been identified by academic science as a key part of ecosystem productivity in the tundra

(Gharajehdaghipour et al. 2016).[16] While we do not know what TEK regarding the fox dens was prior to trade, two facts make it likely that TEK holders were at least partially aware of the fox's place in the ecosystem, in spite of their light use. These are (1) that fox dens are so vibrantly differentiated from the rest of the tundra that they are easily identifiable, even from the air (Garrott et al. 1983), and (2) that when trapping did begin, harvest numbers were high and indicate ease in locating and trapping them (once trap technology arrived via trade).

In reverse, as prices have collapsed for fox furs, a 2011 survey of Canadian Inuit recounts that while 91% of older Inuit hunters (35–49) report that they know where to set fox traps *and why to set them there*, only 46% of younger (18–34) Inuit hold this TEK (Pearce et al. 2011). The return on such TEK has fallen significantly, just as it rose when furs became valuable through trade. The fox fur trade provides an explicit example of increased dispersion in ecosystem use.

This increased dispersion can result in shifts in ecosystem capacity. At the beginnings of the fur trade, the business was so profitable that Russian traders in the west expanded the range of the fox by introducing the species to several of the Aleutian Islands. The absence of other predator mammals and high abundances of birds meant that populations took hold and thrived quickly. Local Inuit were quick to witness declining bird populations to these introductions (Bailey 1993; Croll et al. 2005). While the increase in fox populations increased market profit potential, it did so at the expense of bird populations valuable for subsistence. Now that fur prices have fallen, fox eradication and ecosystem restoration in the Aleutians have increased.

Whether the net impact is positive or negative for society depends on how the ecological interactions translate to well-being. At low levels of extraction, fox trapping increased the dispersion of ecosystem use and allowed the introduction of traded goods and market activity without significantly changing subsistence production and value. However as the fox trade increased, it transferred not only labor and TEK activity to trapping but also ecosystem productivity.

Efficient institutional change takes place when the net benefits to doing so are positive (T. L. Anderson and Hill 1990). In particular, institutions that manage resources through common property, public property, or private property are perceived as alternative solutions to the open access problem, and comparing these institutions according to the extended Demsetz theory (e.g. Witt 1987) involves weighing known enforcement costs against the benefits that a particular institution delivers by reducing free-riding.

In the case of the fur trade, the increased potential for individual returns over collective subsistence investments make system exit a more individual choice – it can happen piecemeal and at the margin instead of as a group decision. System exit would not entail a group movement to new hunting grounds; it rather involves

---

[16]In particular, fox dens promote plant growth that increases nitrogen on the tundra, in turn fostering additional plant growth and animal fodder. This essentially creates garden oases that support and expand the ecosystem's productivity.

individual members of the populations moving in and out of the educational and market opportunities of the larger, wealthy economies in which they became enmeshed (US, Canada, and Denmark) until the TEK becomes so depreciated that return to the system is no longer possible.

Whales and Whaling

Traditional Ecological Knowledge (TEK) as Technology

While we may consider the Arctic as consisting of similar ecological systems, many nuances exist, meaning that TEK pertaining to one species may need refinement or change before being applicable to another, similar-seeming species. Furthermore, with respect to any one species, there may be various methods of harvest.[17]

In addition to changing technology (mainly via trade), different seasons, environmental and geographical conditions allow for different hunting methods and thus productivity change. For example, the presence of shallow water bays increases the productivity of hunting Belugas by driving them into shallow water (Freeman et al. 1998). Also, the productivity of group hunts increases dramatically due to the possibility to employ combinations of hunting techniques. Even though it is possible to hunt Belugas with one or two hunters, i.e. by using harpoons and sealskin floats, larger groups of hunters could work together to drive schools of Belugas ashore and thereby increase the catch (Freeman et al. 1998). Thus, returns to cooperation, TEK, and related management exist.

Depending on culture, geography and the type of whales to hunt, different hunting potentials are present.[18] This also means that TEK is not only specialized on e.g. whaling, storing and sharing, but it is further segregated depending on the geography, climate, season and the type of whale. The combination of those factors shows the variety of circumstances and the complexity of knowing the right thing to do. It determines the productivity of whaling and in turn the potential for development of a managerial (knowledge) group. The ability (and/or desire) to transfer this status into material gain is less clear.

---

[17]The example of Beluga hunting demonstrates many of the hunting techniques and at the same time is an important case in itself, because it is the most hunted whale by Canadian Inuit communities (Freeman et al. 1998). Four main methods exist: the whales can be driven into shallow water, harpooned from the melting and opening ice in spring, shot from ice or shore and shot in open water (Freeman et al. 1998).

[18]One example is the hunting by Inuit in Greenland, where umiaks were used to hunt large, slow swimming whales, like humpback and bowhead whales. Faster swimming whales like fin whales cannot be easily hunted from such boats. Hunting fin whales started with the introduction of motorized boats by Danish colonialist in the in the 1920s (Freeman et al. 1998).

Depending on the whale being hunted, one can define population group sizes that produce more catch, as well as the minimal traditional knowledge of how to operate and manage the hunt.[19] Such type of hunting depends on TEK to make the hunt highly productive. Hunting was productive enough to enable the Inuit expanding along the Canadian coast up to Greenland about 1000 years ago. Harvests could reach up to 60 Bowhead annually along the Alaskan coast (Freeman et al. 1998), and are estimated to have been about 36 bowhead per year amongst Greenlandic and Eastern Canadian Inuit (Higdon 2008). These relatively low numbers in relation to the whale population estimates emphasize the lack of need for governance of the single species and help contextualizes development of governing principles as discussed in e.g. Tyrrell (2007).

Migration patterns of whales make storing and sharing of the whale meat an important issue. Traditional storing knowledge can be viewed as a factor that increased the yield of the whale hunter, while the absence of the knowledge could be regarded as a diminishing factor of the yield. Storing has also the quality to reduce the risk of hunting fewer or no whales, while cultural norms of sharing, even outside the community, can be seen as spreading the risk of hunting fewer whales than needed for population sustenance on a traditional diet.

Whale and Walrus Trade: Depletion and Dispersion

Figure 5 generates specific insight into an Inuit resource system once trade is introduced. The figure graphs the estimated annual whale mortality from the commercial whaling begun in the 1840s in the Pacific Arctic as well as the estimated annual catch of walrus. Once the bowhead was discovered by commercial whalers in 1848, the catch rose immediately, and populations were decimated by the mid-1850s. The Arctic environment made the bowhead whale a great prize for whaling, as it had a much higher blubber and oil content. Its baleen was also considered high quality (Nichols 2009).

In the first years of Pacific Arctic whaling, walrus were primarily ignored, or occasionally harvested for their ivory, which was considered inferior to other types of ivory. The dwindling Bowhead population, however, caused investigation into walrus's oil content and quality, and the resulting success meant that some 150,000 walrus came to be harvested over 65 years.

The first most obvious point is that the combined harvests significantly reduced the ecosystem's productivity for the dependent native population. With no internationally recognized property rights to the marine mammals, no history of overuse

---

[19] This might include hunting with several boats, approaching the slow swimming bowhead whale without disturbing it, while being able to communicate, tire the whale with harpoons and attach floats, finally killing it (Freeman et al. 1998).

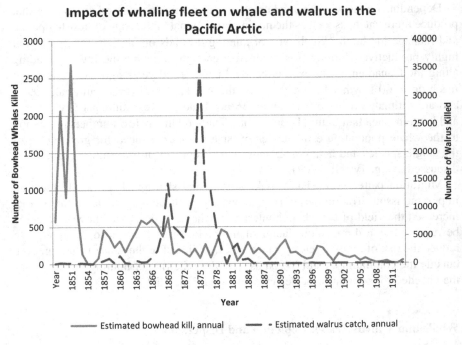

Fig. 5 Whale and Walrus Harvests in the Western Arctic

and related institutional structures suggesting a need for governing the harvest, and no meaningful enforcement tools in any case, the fisheries were open access harvests for the profit of New England.[20]

The data also provides an illustration of how ecosystem knowledge can shift ecosystem use. Walrus are physically easier to harvest than whales due to their behavior on the ice and smaller size. Walrus oil refining costs were lower than whale oil refining costs, and the walrus oil price eventually rose above that of whale oil (Nichols 2009). But learning their commercial value took time.

Further understanding of the consequences of this harvest, and the workings of our model, come from qualitative reports about the visible impacts of the walrus harvest on local Inuit populations. Reports of starvation due to the increasing scarcity of the walrus as early as 1871 suggest wholesale catastrophe lay ahead if action were not taken. These predictions bore out in the winter of 1878–1879 when most of the population of St. Lawrence Island died from starvation. Efforts by at

---

[20]Contrasting this situation was the simultaneous fur seal fishery south of the Bering Strait. The fishery fell under sole authority of the Russian American Company monopoly, and had the important feature that the main seal grounds for reproduction were concentrated on a few uninhabited islands (Pribilof Islands). This made first for quick and profitable overharvest, but upon recognition of the devastation, then allowed a temporary moratorium as early as 1805 to encourage repopulation, followed by limited harvests (Veniaminov 1984).

least one ship captain to stop walrusing were met with agreement that the natives were doomed, but also with the expected tragedy of the commons response where reducing harvest unilaterally simply meant more for another, less cooperative or humane, whaling ship to harvest (Bockstoce 1986; Nichols 2009).

The walrus had only become the staple food for local Inuit after the decimation of these Pacific whales; prior to this, walrus meat was mainly used to feed the dogs. A similar story for the Eastern Arctic is described in Stewart et al. (2014). Here again we see ecosystem dispersion at work.

Conclusions

The long run economic development of the Inuit Arctic presents a set of interesting lessons regarding economic growth through trade, resource governance (and lack thereof), and ecological knowledge and exploitation. The harsh climate of the Arctic and scarcity of resources, particularly for shelter and heat, kept human population growth low enough that resource constraints focused not on overexploitation from tragedy of the commons problems but on preventing underexploitation to achieve survival and subsistence. The importance of TEK in this system cannot be overestimated – without the undoubtedly hard-won wherewithal to turn a few species of large mammals and a handful of other available resources including sea-ice into food, clothing and shelter there would be no Inuit.

This low human population put little pressure on the overall Arctic ecosystem, and required only limited direct governance. Large marine mammal hunting, however, required cooperative action and TEK skills and assets accumulated to successful whaling groups. This ability to transform TEK into subsistence provided only the most limited amounts of stratification, so that Inuit society should be considered extremely egalitarian overall.

This lack of governance needs rendered Inuit societies and institutions unprepared for the advent of new demands on resources from outside sources. Foreign whaling fleets and fur trappers introduced both new technologies and new scarcities. These created multiple impacts on ecosystem use, all with significant consequences for lifestyle. Direct impacts included severe local population loss from starvation when whale and walrus populations were so decimated that a bad weather year could and did result in the deaths of many Inuit from starvation. Indirect impacts have included loss of TEK. This stemmed first from a shift in returns to individual exploitation of ecosystem resources over cooperative harvests of large marine mammals, as fox furs became easily trappable and paid well in currencies that bought new, interesting goods like metal kettles, tobacco, flour and sugar at trading posts. Then, as opportunities shifted further, TEK dissipated due to partial or complete exit from the socio-ecological system into a broader world.

This latter impact is still underway – debates over whale harvesting and seal fur sales continue to press against the ability of Inuit to bridge traditional subsistence activities and modern market activities. Underlying these debates are fundamental questions of the terms of trade for ecosystem services that traditionally supported small Inuit populations, and are now under greater stress from conflicting demands across a multiplicity of goals. Exit from an Inuit socio-ecological system removes extractive subsistence goals from this set of challenges, and as such may have been welcome to many other constituencies over the centuries. At the same time, the lack of governance and property rights concepts that evolved with Inuit resource demands made such exit a relatively simple political and social affair. Today, a greater awareness of the technological ingenuity of Arctic Inuit peoples and their connection to the resource base and its flow of ecosystem services is growing. In this awareness are paths forward for adapting to the planet's increasing resource scarcities and generating the types of capital that create solutions to scarcity as opposed to the types of technology that result in more rapid, more devastating harvest and overuse of natural resources and surrounding ecosystems.

References

Anderson, J. (2011). From Alaska to Greenland: A comparison of the Arctic small tool and Thule traditions. *Totem: The University of Western Ontario Journal of Anthropology, 12*(1), 6.

Anderson, T. L., & Hill, P. J. (1990). The race for property rights. *Journal of Law and Economics, 33*, 117.

Aporta, C. (2009). The trail as home: Inuit and their pan-Arctic network of routes. *Human Ecology, 37*(2), 131–146.

Bailey, E. P. (1993). *Introduction of foxes to Alaskan Islands – History, effects on Avifauna, and eradication.* (93). Homer.

Betts, M. W. (2007). The Mackenzie Inuit whale bone industry: Raw material, tool manufacture, scheduling, and trade. *Arctic, 60*(2), 129–144.

Betts, M. W., & Friesen, T. M. (2004). Quantifying hunter-gatherer intensification: A zooarchaeological case study from Arctic Canada. *Journal of Anthropological Archaeology, 23*(4), 357–384.

Bockstoce, J. R. (1986). *Whales, ice, and men: The history of whaling in the Western Arctic.* Seattle: University of Washington Press. 400 p.

Bockstoce, J. R. (2009). *Furs and frontiers on the far north: The contest among native and foreign nations for the Bering Strait fur trade.* New Haven: Yale University Press.

Brander, J. A., & Taylor, M. S. (1998). The simple economics of Easter Island: A Ricardo-Malthus model of renewable resource use. *American Economic Review, 88*(1), 119–138.

Carboni, O. A., & Russu, P. (2013). Linear production function, externalities and indeterminacy in a capital-resource growth model. *Journal of Mathematical Economics, 49*(5), 6.

Clark, C. (2005). *Mathematical bioeconomics: Optimal management of renewable resources* (2nd ed.). Hoboken: Wiley-Interscience.

Clark, C., & Munro, G. (1975). The economics of fishing and modern capital theory: A simplified approach. *Journal of Environmental Economics and Management, 2*, 92–106.

Coltrain, J. B. (2009). Sealing, whaling and caribou revisited: additional insights from the skeletal isotope chemistry of eastern Arctic foragers. *Journal of Archaeological Science, 36*(3), 764–775.

Coltrain, J. B., Hayes, M. G., & O'Rourke, D. H. (2004). Sealing, whaling and caribou: the skeletal isotope chemistry of Eastern Arctic foragers. *Journal of Archaeological Science, 31*(1), 39–57.

Cooper, H. K., Mason, O. K., Mair, V., Hoffecker, J. F., & Speakman, R. J. (2016). Evidence of Eurasian metal alloys on the Alaskan coast in prehistory. *Journal of Archaeological Science, 74*, 176–183.

Costanza, R., Wainger, L., Folke, C., & Mäler, K. G. (1993). Modeling complex ecological economic systems. *Bioscience, 43*(8), 545–555.

Costanza, R., Cumberland, J. H., Daly, H., Goodland, R., Norgaard, R. B., Kubiszewski, I., & Franco, C. (2014). *An introduction to ecological economics*. Raton: CRC Press.

Croll, D. A., Maron, J. L., Estes, J. A., Danner, E. M., & Byrd, G. V. (2005). Introduced predators transform subarctic Islands from grassland to Tundra. *Science, 307*, 3. https://doi.org/10.1126/science.1108485.

Damas, D. (2002). *Arctic migrants/Arctic villagers: The transformation of Inuit settlement in the central Arctic*. Montreal: McGill-Queen's Press-MQUP.

Emerton, L. (2016). Economic valuation of wetlands: Total economic value. In *The wetland book* (pp. 1–6). Dordrecht: Springer.

Eliasson, L., & Turnovsky, S. (2004). Renewable resources in an endogenously growing economy: Balanced growth and transitional dynamics. *Journal of Environmental Economics and Management, 48*(3), 31.

Field, B. C. (1989). The evolution of property rights. *Kyklos, 42*(3), 319–345.

Finney, B. P., Gregory-Eaves, I., Douglas, M. S., & Smol, J. P. (2002). Fisheries productivity in the northeastern Pacific Ocean over the past 2,200 years. *Nature, 416*(6882), 729–733.

Fisk, E. K., & Shand, R. T. (1970). The early stages of development in a primitive economy: Evolution from subsistence to trade and specialization. In C. R. Walton (Ed.), *Subsistence agriculture and economic development*.

Ford, J. D., Smit, B., Wandel, J., & MacDonald, J. (2006). Vulnerability to climate change in Igloolik, Nunavut: What we can learn from the past and present. *Polar Record, 42*(2), 127–138.

Ford, J. D., Smit, B., Wandel, J., Allurut, M., Shappa, K., Ittusarjuat, H., & Qrunnut, K. (2008). Climate change in the Arctic: Current and future vulnerability in two Inuit communities in Canada. *The Geographical Journal, 174*(1), 45–62.

Freeman, M. M. R., Bogoslovskaya, L., Caulfield, R. A., Egede, I., Krupnik, I. I., & Stevenson, M. G. (1998). *Inuit, whaling, and sustainability*. Walnut Creek: Altamira Press.

Garrott, R. A., Eberhardt, L. E., & Hanson, W. C. (1983). Arctic fox den identification and characteristics in northern Alaska. *Canadian Journal of Zoology, 61*(2), 423–426.

Gharajehdaghipour, T., Roth, J. D., Fafard, P. M., & Markham, J. H. (2016). Arctic foxes as ecosystem engineers: Increased soil nutrients lead to increased plant productivity on fox dens. *Scientific Reports, 6*. https://doi.org/10.1038/srep24020.

Glaeser, E. L., & Shleifer, A. (2003). The rise of the regulatory state. *Journal of Economic Literature, 41*(2), 401–425.

Hardin, G. (1968). The tragedy of the commons. *Science, 162*(3859), 1243–1248.

Higdon, J. (2008). *Commercial and subsistence harvests of bowhead whales (Balaena mysticetus) in eastern Canada and West Greenland*. Ottawa: Fisheries and Oceans Canada.

Kaiser, B. A., & Roumasset, J. (2014). Transitional forces in a resource based economy: Phases of economic and institutional development in Hawaii. *Review of Economics and Institutions, 5*(2), 44.

Kintisch, E. (2016). The lost Norse. *Science, 354*(6313), 696–701.

Laugrand, F., & Oosten, J. (2008). When toys and ornaments come into play: The transformative power of miniatures in Canadian Inuit cosmology. *Museum Anthropology, 31*(2), 69–84.

Lopez, R., Anriquez, G., & Gulati, S. (2007). Structural change and sustainable development. *Journal of Environmental Economics and Management, 53*(3), 15.

McGhee, R. (2007). *The last imaginary place: A human history of the Arctic world*. Oxford: Oxford University Press.

McGhee, R. (2017). The archaeological construction of aboriginality. In C. Hillerdal, A. Karlström, & C.-G. Ojala (Eds.), *Archaeologies of "Us" and "Them": Debating history, heritage and indigeneity*. Florence: Routledge.

Ménard, C., & Shirley, M. M. (Eds.). (2005). *Handbook of new institutional economics* (Vol. 9). Dordrecht: Springer.

Moltke, I., Fumagalli, M., Korneliussen, T. S., Crawford, J. E., Bjerregaard, P., Jørgensen, M. E., et al. (2015). Uncovering the genetic history of the present-day Greenlandic population. *The American Journal of Human Genetics, 96*(1), 54–69.

Morey, D. F., & Aaris-Sørensen, K. (2002). Paleoeskimo dogs of the eastern Arctic. *Arctic, 55*, 44–56.

Nichols, P. (2009). *Final voyage: A story of arctic disaster and one fateful whaling season*. Penguin Group US.

Nuttall, M. (2005). Inuit. In M. Nuttall (Ed.), *Encyclopedia of the Arctic*. London: Routledge.

Pearce, T., Wright, H., Notaina, R., Kudlak, A., Smit, B., Ford, J. D., & Furgal, C. (2011). Transmission of environmental knowledge and land skills among Inuit men in Ulukhaktok, northwest Territories, Canada. *Human Ecology, 39*(271–288). https://doi.org/10.1007/s10745-011-9403-1.

Poncins, G. (1941). *Kabloona*. New York: Reynal.

Raghavan, M., DeGiorgio, M., Albrechtsen, A., Moltke, I., Skoglund, P., Korneliussen, T. S., et al. (2014). The genetic prehistory of the new world Arctic. *Science, 345*(6200), 1255832.

Rasic, J. T. (2016). Archaeological evidence for transport, trade, and exchange in the north American Arctic. In T. M. Friesen & O. K. Mason (Eds.), *The Oxford handbook of the prehistoric Arctic*. Oxford: Oxford University Press.

Ryan, K. (2011). Comments on Coltrain et al., Journal of Archaeological Science 31, 2004 "Sealing, whaling and caribou: the skeletal isotope chemistry of eastern Arctic foragers", and Coltrain, Journal of Archaeological Science 36, 2009 "Sealing, whaling and caribou revisited: Additional insights from the skeletal isotope chemistry of eastern Arctic foragers". *Journal of Archaeological Science, 38*(10), 2858–2865.

Squires, D., & Vestergaard, N. (2013). Technical change in fisheries. *Marine Policy, 42*, 6.

Stern, P. R. (2013). *Historical dictionary of the Inuit*. Lanham: Scarecrow Press.

Stern, H. L. (2016). Polar maps: Captain Cook and the earliest historical charts of the Ice Edge in the Chukchi Sea. *Polar Geography, 39*(4), 220–227.

Stewart, D. B., Higdon, J., Reeves, R. R., & Stewart, E. A. (2014). A catch history for Atlantic walruses (Odobenus rosmarus rosmarus) in the eastern Canadian Arctic. *NAMMCO Scientific Publications, 9*, 219–313.

Taylor, M. S. (2011). Buffalo hunt: International trade and the virtual extinction of the American Bison. *American Economic Review, 101*(7), 3162–3195.

Tyrrell, M. (2007). Sentient beings and wildlife resources: Inuit, beluga whales and management regimes in the Canadian Arctic. *Human Ecology, 35*(5), 575.

Van den Bergh, J. C. (Ed.). (2002). *Handbook of environmental and resource economics*. Cheltenham: Edward Elgar Publishing.

Vaughn, R. (1994). *The Arctic: A history*. Phoenix Mill: Sutton Publishing.

Veniaminov, I. (1984). *Notes on the Islands of the Unalashka District* (L. T. Black & R. H. Geoghegan, Trans. R. A. Pierce Ed.).

Wenzel, G. W. (2009). Canadian Inuit subsistence and ecological instability – If the climate changes, must the Inuit? *Polar Research, 28*(1), 89–99.

Witt, U. (1987). The Demsetz hypothesis on the emergence of property rights reconsidered. In R. Pethig & U. Schlieper (Eds.), *Efficiency, institutions, and economic policy*. Heidelberg: Springer-Verlag.

Zeppel, H. (2006). *Indigenous Ecotourism: Sustainable development and management*. Oxfordshire: CAB International.

Ballast Water and Invasive Species in the Arctic

Henrik Holbech and Knud Ladegaard Pedersen

Abstract Ship's ballast water has been a vector for the spreading of nonindigenous invasive species (NIS) around the globe for more than a century and has had devastating impact on aquatic ecosystems in many regions. Due to the harsh climate, shipping activities in Arctic waters have been limited compared to many parts of the world but will increase in the coming years due to climate changes. This will potentially affect the pristine Arctic marine ecosystems by introduction of NIS. In this chapter, we present the international ballast water regulations that have entered into force and the specific challenges of ballast water management in relation to the Arctic environment and marine ecosystems. We discuss the risk of NIS affecting the Arctic marine ecosystems including the impact of increased shipping activity, changes in living conditions of marine organisms because of climate changes and lack of knowledge of the eco-physiological boundaries and distributions of Arctic marine species. It is concluded that at present only a few marine NIS have been recorded in the Arctic area. Despite the existing and planned ballast water regulations, NIS establishment in the region will increase with an unknown magnitude due to lack of biological data.

Keywords Ship's ballast water • Ballast water regulations • Nonindigenous invasive species • Climate changes • Arctic marine ecosystems • IMO • HELCOM/OSPAR • US Coast Guard

Introduction

It has been estimated that around 5% of the world annual economy is lost due to negative impact of invasive species of plants and animals transported to new ecosystems via anthropogenic activities (Pimentel et al. 2001). A part of this loss is caused by aquatic invasive species spread by shipping activities including release of ships ballast water between regions and ecosystems. This chapter deals

H. Holbech (✉) • K.L. Pedersen
Department of Biology, University of Southern Denmark, Odense, Denmark
e-mail: hol@biology.sdu.dk

© Springer International Publishing AG 2018
N. Vestergaard et al. (eds.), *Arctic Marine Resource Governance and Development*,
Springer Polar Sciences, https://doi.org/10.1007/978-3-319-67365-3_7

with the parameters affecting the risk of introducing Nonindigenous Invasive Species (NIS) to the Arctic region through ships ballast water. The parameters are discussed in the light of the newly ratified United Nations International Maritime Organisation (IMO) ballast Water Management Convention (BWM) that entered into force in September 2017 and the US Coast Guard Ballast Water Discharge Standard that entered into force in 2012. The consequences of climate changes including changed living conditions for organisms in Arctic waters and increased anthropogenic activity are discussed in relation to the potential of NIS to affect the Arctic ecosystems. First we summarize the use of water as ballast in ships and give examples of NIS spread by ballast water. The ballast water regulations will be briefly described for selected regions and the risks of NIS introduction to Arctic associated with the current and upcoming ballast water regulations will be discussed. Finally we identify the challenges in predicting the risk for establishment of NIS via ballast water in the Arctic region.

Ballast Water (BW)

The use of ballast to stabilize ships is as old as sailing itself. Until the late nineteenth century ships used solid materials such as rocks or sand as ballast but the risk of relocation of solid ballast during rough weather conditions and long loading times made water preferable as ballast. In around 1880, ship construction and design made it possible to use water as ballast instead of solid materials and for more than 130 years ballast water (BW) has been used to control list, draught, stability or stresses of the ship and to compensate for changes in cargo load levels (IMO 2004, 2016). BW is taken up by pumps or gravity feed via sea chests and typically BW is taken up during cargo unloading and released during cargo loading in ports (Fig. 1). Ships carrying BW usually have a capacity of 20–50% of the ships dead weight tonnage (DWT) dependent on ship type (David et al. 2012). The amount of BW transferred yearly is enormous: It has been estimated that the amount of overseas BW released in the United States coastal waters in 2006–2007 was around 111 million metric tons and so called coastwise or intra-coastal discharges were 280 million metric tons in the same period (Miller et al. 2011).

It is estimated by the International Maritime Organization (IMO 2016) that the yearly transport of BW is around 10 billion tons globally (Awad et al. 2014; Tamelander et al. 2010).

Nonindigenous Invasive Species (NIS)

A negative side-effect of the intensive use of BW is the unintended introduction of nonindigenous invasive species (NIS) to new ecosystems around the globe when vessels take up BW at one site and release it at another site.

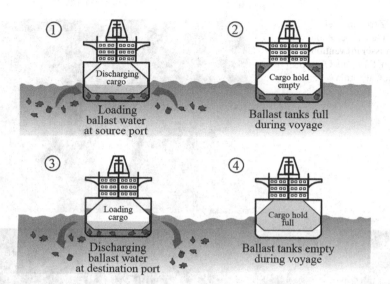

Fig. 1 Principle of BW exchange (https://commons.wikimedia.org/wiki/File:Ballast_water_en.svg)

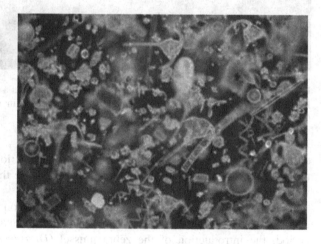

Fig. 2 The organism content in BW is extremely complex and consists of thousands of animal and plant species including algae and zooplankton as well as fungi, bacteria and viruses (http://orma.com/sea-life/plankton-facts/)

According to the International Maritime Organization (IMO) the spread of invasive species is recognized as one of the greatest threats to the ecological- and economic wellbeing of the planet (IMO 2016). It is estimated that thousands of species are regularly (daily) transported in the biological complex BW (Fig. 2) (Committee on Ships' Ballast Operations 1996) and aquatic NIS are already affecting aquatic ecosystems worldwide. In 2006 more than 1000 NIS had been registered in European waters alone (Gollasch 2006). BW accounted for 22% and hull fouling for 16% of the European NIS registered in 2006. BW is normally one of the main contributors to aquatic NIS (DiBacco et al. 2012) but in the European

Fig. 3 Zebra mussels attached to a pipe (Photo from Mussel Prevention Program, San Luis Obispo County, California)

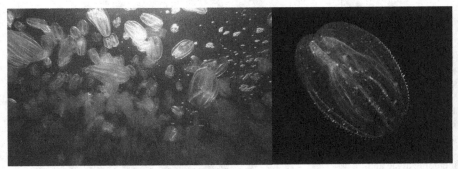

Fig. 4 The invasive Sea Walnut (*Mnemiopsis leidyi*) was a part of a crash in the ecosystems in the Black Sea and Caspian Sea and caused a nearly 100% reduction in anchovy fisheries in the 1990s (Credit: Marco Faasse, World Register of Marine Species)

case shipping was not the only vector for NIS introduction because the opening of the Suez Canal caused the direct introduction of more than 250 NIS to European waters (Gollasch 2006).

Many NIS have already had a devastating impact on ecosystems and regional economies: In 2009 the U.S. Department of State estimated that within a 10 year period, the introduction of the zebra mussel (*Dreissena polymorpha*) from the Caspian Sea and the Black Sea in Europe to the Great Lakes, would cost more than 3 billion dollars due to water pipe clogging (Fig. 3) and that the zebra mussel will outcompete around 50% of native mussels causing the extinction of up to 140 species (Stein and Flack 1996).

In 1982, The comb jellyfish *Mnemiopsis leidyi,* also called sea walnut, was introduced to the Black Sea by ships BW from the Atlantic coast of North America (Fig. 4). The jellyfish had no native enemies and preyed on zooplankton which caused a crash of the ecosystems. Within a few years it accounted for 90% of the total biomass in the Black Sea. The main reason was probably overfishing stocks normally preying on zooplankton – thereby increasing the zooplankton abundancy and creating favorable conditions for the explosion in *M. leidyi* biomass (Gucu 2002). The jellyfish grazed the major part of the zooplankton and also fish larvae and the fisheries collapsed. A later non-intended introduction of another jellyfish species (*Beroe ovata*) predating on *M. leidyi* has reduced the problem and an increase in fish stocks has been observed the last years.

Ballast Water Regulations

To protect aquatic ecosystems from further introduction of NIS, international and national initiatives have been developed during the last decades. A recent publication reviews the main BW management policies developed since the 1990s with focus on Alaska (Verna and Harris 2016). This section focuses on international and regional BW regulations with impact on the Arctic region.

The International Maritime Organisation (IMO)

> **Info-box 1: The IMO BWM Convention (Adopted in 2004, entered into force in 2017)**
>
> The BWM convention has the purpose to "*prevent the spread of harmful aquatic organisms from one region to another, by establishing standards and procedures for the management and control of ships' ballast water and sediments*"
>
> *Ballast Water Management means mechanical, physical, chemical, and biological processes, either singularly or in combination, to remove, render harmless, or avoid the uptake or discharge of Harmful Aquatic Organisms and Pathogens within Ballast Water and Sediments*
>
> (IMO 2004)

In 2004, IMO adopted the International Convention for the Control and Management of Ships' BW and Sediments (BWM) after more than 14 years of negotiations between member states (IMO 2004). The IMO BWM will enter into force 12 months after ratification by 30 States, representing 35 per cent of world merchant shipping tonnage. This goal was achieved in September 2016 when Finland ratified the convention. The IMO BWM entered into force in September 2017 (IMO 2016). The BWM requires that all ships discharging BW must first apply a type approved BW management system (BWMS) to meet discharge standards related to the number of viable organisms in defined size-classes. The BWM can roughly be described as a two phased strategy with Regulation D-1 (Info-box 2) explaining how to perform a mid-ocean ballast water exchange according to specific standards and Regulation D-2 (Info-box 3) setting up the specific standards on how to comply with ballast water management by using a ballast water management system. Regulation D-1 is a transient regulation that will be replaced by Regulation D-2 within specified time frames that depends on the age of the vessel and the amount of ballast water transported by the vessel.

There are a number of conditions described in Regulation A-3 and A-4 under which exemptions may be granted to apply with Regulation D-2. These exemptions can be considered as potential gaps in the prevention of spreading NIS, The IMO Exemptions are included in Info-box 4 and also discussed in the HELCOM/OSPAR section. This aspect will be discussed in relation to the Arctic and the USCG regulation in the section on risk of NIS in the Arctic region.

After the adoption of the BWM, 14 Technical Guidelines have been developed and published to support implementation of the BWM (IMO 2008). The BWM Regulation D-2 is defining the maximum amount of viable organisms in specified size classes that is accepted to be released with BW. The details are described in Info-box 3.

Info-box 2: IMO BWM Regulation D-1: Ballast Water Exchange Standard

1 Ships performing Ballast Water exchange in accordance with this regulation shall do so with an efficiency of at least 95 percent volumetric exchange of Ballast Water.

2 For ships exchanging Ballast Water by the pumping-through method, pumping through three times the volume of each Ballast Water tank shall be considered to meet the standard described in paragraph 1. Pumping through less than three times the volume may be accepted provided the ship can demonstrate that at least 95 percent volumetric exchange is met.

(IMO 2004)

Info-box 3: IMO BWM Regulation D-2: Ballast Water Performance Standard

1 Ships Conducting Ballast Water Management in Accordance with this Regulation shall Discharge Less than:

10 viable organisms per m3 greater than or equal to 50 μm in minimum dimension.

and less than 10 viable organisms per ml less than 50 μm in minimum dimension and greater than or equal to 10 μm in minimum dimension;

And discharge of the indicator microbes shall not exceed the specified concentrations described in paragraph 2.

(continued)

2 Less than the following concentrations of indicator microbes, as a human health standard:

1. *Toxigenic Vibrio cholerae (O1 and O139) with less than 1 colony forming unit (cfu) per 100 ml or less than 1 cfu per 1 g (wet weight) zooplankton samples.*
2. *Escherichia coli less than 250 cfu per 100 ml.*
3. *Intestinal Enterococci less than 100 cfu per 100 ml.*

(IMO 2004)

United States (U.S.)

Based on evidence that NIS such as the zebra mussel were released into the Great Lakes via ballast water, the United States Congress approved the Non-indigenous Aquatic Nuisance Prevention and Control Act of 1990 (NANPCA) – a federal program that directed the U.S. Coast Guard to issue regulations to prevent the introduction and spread of aquatic NIS into the Great Lakes through ballast water (Congress 1990).

An amendment to NANPCA came in 1996 with The National Invasive Species Act of 1996 (NISA), which expanded ballast water management to all U.S. waters (Congress 1996). Among other initiatives NISA directed guidance on mid-ocean ballast water exchange practices.

In June 2012, the final U.S. Coast Guard BW Discharge Standard entered into force and set up regulations of BW discharge in United States waters. It applies to the U.S. territorial sea, or out to 12 nautical miles, and concerns both sea going vessels and coastwise vessels above 1600 Gross Tonnage (GT) operating between so called Captain of the Port Zones (Fig. 5) that are defined geographical zones (USCG 2012). The discharge standard aligns with the IMO BWM regulation D-2 (Info-box 3).

Canada

The current Canadian Ballast Water Control and Management Regulations apply to Canadian vessels everywhere and foreign vessels in waters under Canadian jurisdiction (Canada 2011). Both ballast water exchange and ballast water management are accepted and generally the requirements are in line with the IMO BWM. For vessels navigating transoceanic routes, rules are set to exchange ballast water at a depth of at least 2000 meters and more than 200 nautical miles from Canadian shore. If vessels

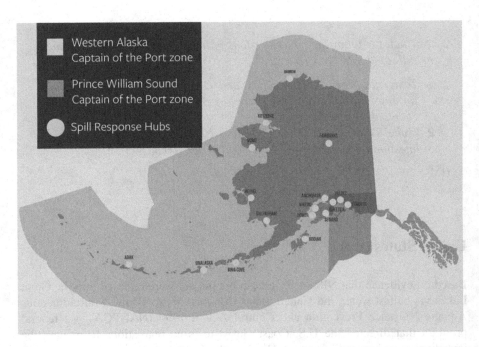

Fig. 5 Captain of the Port Zone for Western Alaska and Prince William Sound (Alaska Maritime Prevention & Response) (From http://www.ak-mprn.org/)

do not navigate 200 nautical miles from shore en-route, exchange must take place at least 50 nautical miles from shore at a minimum depth of 500 meters. Vessels using a ballast water treatment system must adhere to the same standards for maximum numbers of viable organisms as described in IMO BWM Regulation D-2 (IMO 2004). A number of exemptions are also described in the Canadian regulation and will be discussed in the risk of NIS section.

HELCOM/OSPAR Member Parties

HELCOM (the Baltic Marine Environment Protection Commission – Helsinki Commission) is the governing body of the Helsinki Convention on the Protection of the Marine Environment of the Baltic Sea Area (HELCOM 2016). The OSPAR commission was set up by the 1992 OSPAR Convention for the Protection of the Marine Environment of the North-East Atlantic (OSPAR 2016). Member parties include states around the Baltic Sea, the European Union and countries important in relation to the Arctic region such as Russia, Norway, Finland, Sweden, Denmark, Iceland and the UK. The HELCOM/OSPAR members are following the IMO BWM regulation, but to harmonize exemptions to comply with Regulation D-2, the HELCOM and OSPAR Commissions have agreed on the "*Joint Harmonized*

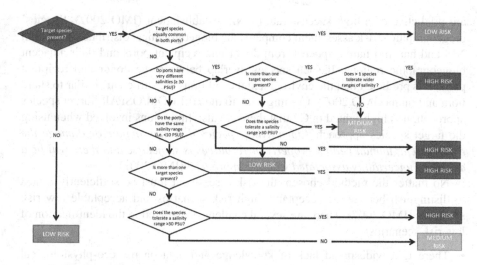

Fig. 6 HELCOM/OSPAR risk assessment algorithm for defining risk of NIS introduction between defined ports within the HELCOM/OSPAR regions

Procedure for the Contracting Parties of OSPAR and HELCOM on the granting of exemptions under the International Convention for the Control and Management of Ships' Ballast Water and Sediments, Regulation A-4." This is to ensure that exemptions are granted in a constant manner that prevents damage to the environment, human health, property or resources, once the BWM Convention enters into force (OSPAR/HELCOM 2015). Therefore a joint task Group on Ballast Water Management Convention Exemptions (HELCOM/OSPAR TG BALLAST) has been established in 2013 and is working on developing a common framework on the specific issue of exemptions for the IMO BWM (IMO 2004) for the Baltic Sea and the North-East Atlantic regions (HELCOM 2016; OSPAR 2016).

A group of NIS among all NIS present in the HELCOM/OSPAR area has been identified as of particular relevance in the context of the IMO BWM requirement. This group of species is the HELCOM/OSPAR Target species, e.g., which are species in other regions that may impair or damage the environment, human health, property and resources in the HELCOM/OSPAR area. Figure 6 presents the risk assessment algorithm using the target species as starting point to evaluate the risk of introducing NIS between ports.

The use of target species – also called Species-specific risk assessment is one of three methods outlined in the guidelines for risk assessment under regulation A-4 of the IMO BWM convention – G7 (IMO 2007). The other two methods are Environmental matching risk assessment and Species' biogeographical risk assessment. Environmental matching risk assessment compares the environmental conditions i.e. salinity and temperature between the donor region where ballast water is taken up with the recipient port where it is released. Environmental matching risk assessments have limited value where the differences between a donor biogeographic region and a recipient port are small because similar environments are predictive of a high success rate of NIS establishment (IMO 2007). Species' biogeographical risk assessment compares the biogeographical distributions of both NIS and harmful native species from donor and recipient ports and their adjacent biogeographic regions. If such species overlap between the donor and recipient ports, it is predicted that the environmental conditions are sufficient similar to share flora and fauna (IMO 2007). Coming back to the HELCOM/OSPAR Target species approach, it is highlighted in G7 that there are also limitations involved when using the target species approach: *"identifying species that may impair or damage the environment, human health, property or resources is subjective and there will be a degree of uncertainty associated with the approach"* (IMO 2007).

No matter the method chosen, the risk assessment must be sufficiently robust to distinguish between unacceptable high risk scenarios and acceptable low risk scenarios (IMO 2007). There are several challenges involved in the identification of low risk scenarios:

- There is a widespread lack of knowledge and data on the eco-physiological capacities of species i.e., which physical and chemical conditions set the limits for successful reproduction. For example Ware et al. (2016) only found sufficient data for about 35% of the species this study investigated.
- The mechanism behind the establishment of new species in an ecosystem and the transition to NIS is poorly understood. Sometimes a so-called lag time of several decades is observed from the first introduction of a species before significant population growth turn it into a NIS (Mack et al. 2000). In other cases the new species start to grow in population size from the point of entrance as seems to be the case for both the zebra mussel in the Great lakes and the comb jellyfish in the Black sea.
- Systematic monitoring of species composition in ecosystems is required to avoid new potential NIS introduced to a donor port after a risk assessment, are spread throughout an ecological zone.
- The risk assessment approaches are further challenged by the fact that more or less all knowledge about NIS population dynamics is based on retrospective observations and not on predictions, which is actually what is done during a risk assessment.
- To be able to compare ecosystems and define ecological zones as needed in Species' biogeographical risk assessment it is necessary to know the species of the ecosystems and their distribution within the systems. For many of the Arctic systems such data are scarce.

Climate Changes in the Arctic Region

Sea water temperatures are increasing globally due to climate changes and in the Arctic the increase is predicted to be larger than in other regions (IPCC 2013; Philippart et al. 2011; Rhein et al. 2013). Observed rates of increase in surface temperature in the Arctic have been twice that of the rest of the world. This has

Table 1 Ten lowest minimum Arctic sea ice extents. Satellite record, 1979 to 2016 (NSIDC 2016)

Rank	Year	10^6 km^2	Date
1	2012	3.39	Sept. 17
2	2016	4.14	Sept. 10
	2007	4.15	Sept. 18
3	2011	4.34	Sept. 11
4	2015	4.43	Sept. 09
5	2008	4.59	Sept. 20
6	2010	4.62	Sept. 21
7	2014	5.03	Sept. 17
8	2013	5.06	Sept. 13
9	2009	5.12	Sept. 13
10	2005	5.32	Sept. 22

resulted in dramatic reductions of sea-ice coverage at its minimum in September, but also in the thickness of the winter ice with a maximum in March.

In 1978, satellites began monitoring sea ice in the Arctic. Using the average sea-ice cover in September for the period 1979–2000 as a reference point (7 million square kilometers) the cover in record breaking 2012 had declined by 50% (Table 1). The decline is accelerating due to the positive feedback phenomenon caused by the increasing absorption of solar energy by the ever-darker surface of the ocean (Walsh 2013). Consequently, estimates of the decline in ice cover per decade have constantly been up-regulated, from 6.5% in 2001, 8.5% in 2005, 10.2% in 2007 to 12% decline per decade in 2011 (Maslowski et al. 2012). Apparently, IPCC models for Arctic ice cover have been too conservative and after updating with the new observations it has been estimated that the Arctic ocean will be ice free in September some time between 2028 and 2037 (Wang and Overland 2009).

The area of the ice cover during the Arctic winter has not shown the same dramatic decline with rates of only a few percent per decade being recorded. However, the thickness of the March ice cover has been reduced by 1.8 m in the period between 1978 and 2008 and climate models predict a reduction from the 2.5 m at present to only 1.2 m when the Arctic ocean is ice free in the summer (IPCC 2013).

Besides declining ice-coverage, water temperature and salinity are expected to change over the next decades (Rhein et al. 2013).

Effects of Climate Changes on the Threat from Invasive Species in the Arctic

The predicted climate changes in the Arctic will greatly enhance the risk of introducing new species to the vulnerable ecosystems. Changed water temperature and salinity are both parameters expected to impact the risk of NIS to be "successfully" introduced to the Arctic regions because many more species are naturally adapted to lower salinity and higher temperature than what has been the norm in the Arctic. The main vector for introduction of NIS is shipping and since ship transport is expected to increase sharply as sea-ice decreases the risk of releasing NIS from ballast water or hull fouling will increase. The NIS risk associated with ballast water is complex and relates not just to the type and amount of ballast water but also the sailing route and time between ballast water uptake and release as well as the similarity of the ecosystems of the uptake- and release points respectively. Further, climate change derived physico-chemical factors like increase in sea temperature, decrease in coastal salinity and shortening of oxygen free periods during winter will all contribute to an environment where many more species are able to survive and reproduce (Jing et al. 2012). The changes in shipping activity and in living conditions for marine species are described in the two following sections.

Increased Shipping Activity in the Arctic

In September 2013 a commercial ship transited the Northwest Passage (NWP) (see Fig. 7) for the first time thereby testifying that climate changes in the Arctic are not a theoretical outcome of models but today's reality. The same year a peak of 71 ships transited through the Arctic routes, a number that has since declined (Miller and Ruiz 2014).

However, despite the marked increase in transiting vessels through the NWP and NSR these routes will probably not be significant alternatives to shipping through the Suez Canal in the near future. A recent publication has investigated the economic feasibility of shipping across the Arctic sea routes like the NSR in a quantitative study covering the next 35 years (Hansen et al. 2016). The authors estimated that large scale liner shipping to the American West coast and Asia via Arctic routes will only become economically attractive around year 2040.

Much more important for the increasing amount of shipping now and in the near future is the traffic to and within the polar region aimed at extracting local resources. In a quantitative assessment of Arctic shipping between 2010 and 2014, Eguiluz et al. (2016) used data from the Automatic Identification System to map the presence of ships in different categories (Fig. 8). In 2014 a total of 11,066 ships were detected, the majority being research-, supply- and survey vessels ('Other' category). This was followed by fishing (1960), cargo (1892), tanker (524) and passenger (308) vessels. A significant new contribution to Arctic shipping has come from tourist cruises, which has been increasing steadily for the last two decades. In 2011 more than 40,000 cruise passengers aboard 35 ships visited Svalbard while 30,000 passengers visited Greenland. In the Canadian Arctic the cruising activity has stabilized at around 20 annual voyages (Hansen et al. 2016).

It has been suggested that tourism trends, commodity prices and natural resource development were more important drivers of increased shipping activity in the Arctic than the amount of ice coverage and hence climate changes (i.e. (Dawson et al. 2014; Eguiluz et al. 2016). However, a recent analysis of shipping in

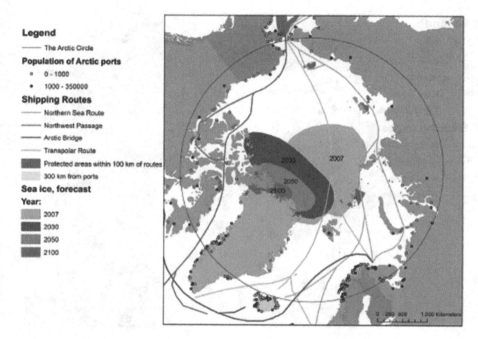

Fig. 7 Future Arctic shipping routes mapped in relation to protected areas, distance from ports and a forecast of the reduction of the Arctic sea ice (Eliasson et al. 2017)

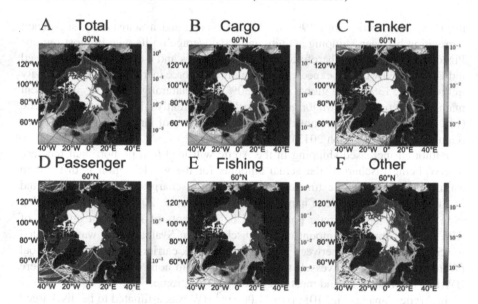

Fig. 8 Average monthly densities of ships in the Arctic in 2014 (Eguiluz et al. 2016)

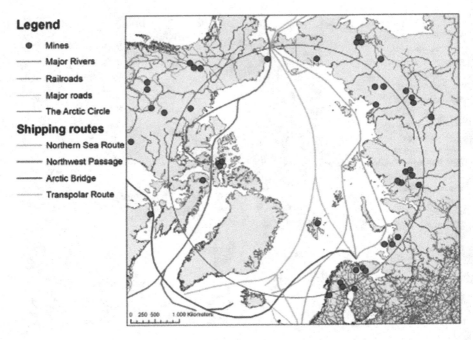

Fig. 9 Mines in the Arctic region and the potential for transporting minerals (Eliasson et al. 2017)

the Canadian Arctic from 1990 to 2015 demonstrated a statistically significant correlation between shipping activity and ice coverage on its own in most regions Pizzolato et al. 2016). Therefore the climate changes and a considerably increased industrial activity can be expected to mutually re-inforce increased shipping activity in the Arctic. 13% of the global oil and 30% of the natural gas reserves are present mainly in Arctic Russia. Major mining activity is currently taking place in the Arctic (Fig. 9) and Greenland is considering the exploitation of large rare-metal reserves (Gautier et al. 2009; Toph 2013).

Although increased shipping in the Arctic will lead to a proportional increase in BW being discharged the actual amounts for the whole region is difficult to estimate. However, two estimates for the present discharges in Arctic Canada and Svalbard have been made (Chan et al. 2013; Ware et al. 2014). It was shown that even relatively small Arctic ports could be recipients of significant amounts of BW. In 2011 the total amount of BW discharged in Svalbard port was estimated to 653,000 m^3 divided between 31 ships of the bulk carrier type transporting coal (Ware et al. 2014). The vessels typically visited from non-Norwegian ports where BW had been taken up and most often exchanged en route.

In Arctic Canada (Fig. 10) up to 41,000 m^3 BW was estimated to be discharged annually in Churchill (Chan et al. 2013).

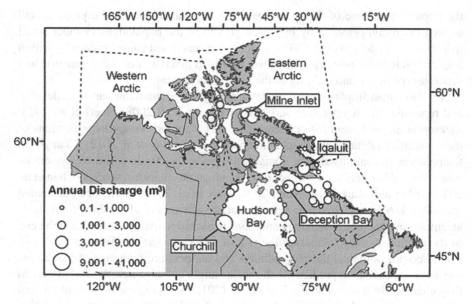

Fig. 10 Corrected ballast water discharges in Canadian Arctic ports (Chan et al. 2013)

Changes in Living Conditions of Marine Organisms

The Arctic marine ecosystem is still relatively unaffected by NIS compared to temperate and tropic regions because shipping activity has been low due to extreme weather conditions and ice coverage. An example of this is that in European Arctic waters, only 18 NIS were registered in 2006 where the number for the Mediterranean was more than 300 (Gollasch 2006).

The complexity of ecosystems and species diversity is negatively correlated with latitude (Rex et al. 2000) although species diversity patterns in Arctic waters have been shown to be unexpectedly complex (Yasuhara et al. 2012). As a less species rich system the Arctic marine ecosystem is therefore predicted to be even more vulnerable to NIS than temperate and tropic marine ecosystems. This is also supported by the diversity-stability hypothesis saying that diversity stabilizes community and ecosystem properties (Tilman 1996). A recent study on the marine ecosystem of California support that species-rich systems are more stable and concludes that so-called functional complementarity is the primary mechanism promoting resilience and long-term stability (Lindegren et al. 2016). Functional complementarity is when species differ in their contribution to a collective function. Lindegren et al. (2016) as well as Chavez et al. (2003) show that functional complementarity is manifested by opposite responses of functionally similar species to positively auto-correlated drivers like temperature and salinity. An example could be that several different small fish species act as prey for larger predators and a larger number of these small species increases the success of the predators. A decrease in the population of one of the species will not be catastrophic for the predator and is even often compensated by increased growth of the populations of other small fish species. Lindegren et al. (2016) suggest that functional complementarity within trophic levels is the primary mechanism by which diversity maintains function and promotes resilience and stability of the marine ecosystem.

The two most important factors determining where plankton can live, develop and reproduce are temperature and salinity (Barry et al. 2008; Floerl et al. 2013; Jackson et al. 2009) and it has been shown that plankton occupy large portions of their potential habitats with suitable temperatures (Sunday et al. 2012). Changes in temperature and salinity during climate changes in the Arctic will therefore create new potential habitats for a number of non-indigenous plankton species (Chan et al. 2013; Miller and Ruiz 2014; Niimi 2004; Ware et al. 2016). It has been estimated that of the 136 ports from where vessels visited Svalbard in 2011, a third would have an environmental match with the conditions predicted to exist in Svalbard at the end of the century thereby increasing the risk of NIS considerably (Ware et al. 2014). From eco-physiological data for salinity- and temperature tolerance of planktonic species it is possible to predict which species might colonize new Arctic habitats as they changes. In a recent study, Ware et al. (2016) analyzed the biological content of ballast water discharged in Svalbard and predicted the possibility of colonization by the introduced species. Sufficient eco-physiological data were available for eight of the 23 identified species and predictions of their colonization were given under present climatic conditions as well as under ocean climate forecasts for 2050 and 2100 using the Representative Concentration Pathway (RCP) 8.5 model. For a single of the species found in the ballast water (*P. leuckartii*) conditions were already suitable for colonization in Svalbard. Under the 2050 scenario five of the eight investigated species could be expected to colonize Svalbard while all might colonize under the 2100 scenario. It should be noted that the actual number of species capable of colonization might be much higher since eco-physiological data for predictions were only sufficient for 8 out of 23 species found in the ballast water (Fig. 11).

Challenges of Ballast Water Management in the Arctic

The IMO BWM and several regional and national regulations and rules concerning discharges of ballast water are being ruled out these years with enormous economic impact on the shipping industry. It is estimated by shipowners that more than 30 billion U.S. dollars are to be invested in installation and maintenance of BWMS affecting more than 60,000 vessels within a 10 year period as a consequence of the IMO BWM and U.S. Coast Guard final decision.

Fig. 11 *Podon leukartii*, a zooplankton organism found in ballast water in Svalbard for which environmental conditions are already suitable for colonization (http://www.cladocera.de/cladocera/taxonomy/pod.html)

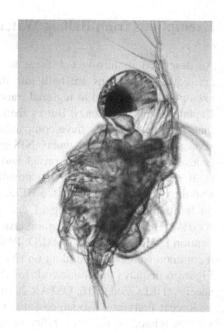

Organisms Below 10 μM in Size

The question is how efficient these regulations will be in preventing NIS to be introduced to the Arctic region. There is no doubt that the BWM regulation D-2 (equal to the U.S. Coast guard discharge standards) when fully implemented will decrease the spreading of NIS above 10 μM in size between different regions. But several species of organisms, especially phytoplankton are smaller than 10 μM and some of them might be more resistant to BWMS than larger organisms. BWMS are at the moment being approved based on compliance with the D-2 requirements.

Ballast Water Exchange

Exchange of Ballast Water at open sea has been the common method to prevent the spread of NIS for the last decades. The IMO D-1 Ballast Water Exchange regulation and other comparable regulations from Canada and the United States describe the accepted exchange methodologies (Info-box 1). It will for some years be allowed for a large number of vessels to follow IMO regulation D-1 mid-ocean ballast water exchange until a BWMS has been installed. Several studies including Ware et al. (2016), have shown that ballast water exchange does not prevent the introduction of NIS to the Arctic region. It is therefore important that the risk of water exchange instead of water management is taken into account especially when Arctic ports are visited in the nearer future.

Exemptions from Ballast Water Regulations

As described in info-box 4 there are several exemptions to D-2 requirements and equal exemptions are built into the Canadian and U.S. regulations. Mostly exemptions are related to regional travelling as for example travelling within U.S. Captain of the Port Zones. But as seen in Fig. 5, such zones can cover large areas that do not necessarily have comparable ecosystems and/or species composition. Also secondary invasion where NIS are spread within regions via ballast water is a concern that has been raised and documented (Chan et al. 2014; DiBacco et al. 2012). The risk for NIS spreading due to the regulative exemptions was also the background for why the HELCOM and OSPAR Commissions have agreed on harmonizing the granting of exemptions under IMO BWM regulation A-4 (OSPAR/HELCOM 2015). A joint task Group on Ballast Water Management Convention Exemptions (HELCOM/OSPAR TG BALLAST) is working on developing a common framework building on risk assessment of NIS invasion between ports. Figure 6 displays the framework for the risk assessment based on selected target species (HELCOM 2016; OSPAR 2016).

Recent literature also suggest that risk assessment based on NIS and different vectors related to the probability of transferring NIS between regions is a way forward to reduce the risk of spreading NIS (Briski 2012; Chan et al. 2013; Claudi and Ravishankar 2006).

Info-box 4: Exemptions in IMO BWM

1 *A Party or Parties, in waters under their jurisdiction, may grant exemptions to any requirements to apply regulations B-3 or C-1, in addition to those exemptions contained elsewhere in this Convention, but only when they are:*

> *1 granted to a ship or ships on a voyage or voyages between specified ports or locations; or to a ship which operates exclusively between specified ports or locations;*
> *2 effective for a period of no more than five years subject to intermediate review;*
> *3 granted to ships that do not mix Ballast Water or Sediments other than between the ports or locations specified in paragraph 1.1; and.*
> *4 granted based on the Guidelines on risk assessment developed by the Organization.*

2 *Exemptions granted pursuant to paragraph 1 shall not be effective until after communication to the Organization and circulation of relevant information to the Parties.*

(continued)

3 Any exemptions granted under this regulation shall not impair or damage the environment, human health, property or resources of adjacent or other States. Any State that the Party determines may be adversely affected shall be consulted, with a view to resolving any identified concerns.

(IMO 2004)

Hull Fouling

Most attention and regulation is focusing on ballast water but hull fouling appears also to be an important vector for introduction of NIS between regions (Chan et al. 2015; Gollasch 2002, 2006). Hull fouling has been a vector for NIS transportation between regions ever since humans started to sail and even long before that by large marine mammals performing annual migration between temperate and Arctic regions (Fig. 12).

Ballast water has only been used for a little more than 100 years (Hewitt and Campbell 2010) and therefore represents a relatively new vector. Ballast water facilitate distribution of pelagic species (or species with pelagic stages) and sediment living species that are not likely to be transported on hulls as well as other life stages of organisms transported on hulls, e.g. planktonic life stages (Hewitt and Campbell 2010). There is recent evidence that both algae and invertebrate organisms transported from temperate to Arctic marine environments on ships hull can survive and potentially become NIS (Chan et al. 2016). So, paying attention to ballast water in this chapter should not prevent focus on also ship hulls as a vector of NIS in the Arctic region.

Fig. 12 Biofouling on the tail of a Humpback whale (©Colourbox.com)

Conclusions

For a century, many species of aquatic organisms have been carried around the globe in ships' ballast water and discharged in foreign ecosystems. Some have settled in their new environment and caused ecological and economical disasters by destroying local food chains and fisheries and by fouling technical equipment and ships. To prevent further loss, IMO's international convention for the control and management of ships' ballast water and sediment was ratified in 2016. Comparable regulations are implemented in the US and Canada. They require that all ships carrying ballast water must have means onboard to prevent the discharge of viable organisms with the ballast water. Until now very few invasive species have settled in Arctic seas due to the extreme physical conditions for life and a limited shipping activity. However, the climate changes experienced in the Arctic region in the last decades have diminished the extent of sea-ice which has made an increase in shipping possible with a concomitant increase in discharged ballast water. Further, the climate changes will improve the chances of survival and settling of new species introduced by ballast water in a highly vulnerable ecosystem. Despite the new regulations there are still challenges to prevent invasive species to settle and spread. First of all, the regulations do not require released ballast water to be 100% free of viable organisms. Therefore, there will always be a risk of introducing NIS even though the regulations are followed. Secondary, because of exemptions to the regulation requirements, there are uncertainties regarding the capability of the regulations in preventing viable organisms to be discharged with ballast water even though risk assessment have been performed according to guidelines from IMO and HELCOM/OSPAR. A major challenge in relation to assessing the risk of introducing NIS to the Arctic via ballast water is the lack of knowledge of the Arctic ecosystem species – including their eco-physiological boundaries and their roles in the system as well as their temporal and spatial distributions.

References

Awad, A., Haag, F., Anil, A. C., & Abdulla, A. (2014). Guidance on port biological baseline surveys. In *GEF-UNDP-IMO GloBallast Partnerships Programme, IOI, CSIR-NIO and IUCN*. GEF-UNDP-IMO GloBallast Partnerships, London, UK. GloBallast Monograph No. 22(22).

Ballast Water Control and Management Regulations (SOR/2011-237), (2011).

Barry, S. C., Hayes, K. R., Hewitt, C. L., Behrens, H. L., Dragsund, E., & Bakke, S. M. (2008). Ballast water risk assessment: Principles, processes, and methods. *ICES Journal of Marine Science, 65*(2), 121–131. https://doi.org/10.1093/icesjms/fsn004.

Briski, E., Ghabooli, S., Bailey, S. A., & MacIsaac, H. J. (2012). Invasion risk posed by macroinvertebrates transported in ships' ballast tanks. *Biological Invasions, 14*(9), 1843–1850. https://doi.org/10.1007/s10530-012-0194-0.

Chan, F. T., Bailey, S. A., Wiley, C. J., & MacIsaac, H. J. (2013). Relative risk assessment for ballast-mediated invasions at Canadian Arctic ports. *Biological Invasions, 15*(2), 295–308. https://doi.org/10.1007/s10530-012-0284-z.

Chan, F. T., Briski, E., Bailey, S. A., & MacIsaac, H. J. (2014). Richness-abundance relationships for zooplankton in ballast water: Temperate versus Arctic comparisons. *ICES Journal of Marine Science, 71*(7), 1876–1884. https://doi.org/10.1093/icesjms/fsu020.

Chan, F. T., MacIsaac, H. J., & Bailey, S. A. (2015). Relative importance of vessel hull fouling and ballast water as transport vectors of nonindigenous species to the Canadian Arctic. *Canadian Journal of Fisheries and Aquatic Sciences, 72*(8), 1230–1242. https://doi.org/10.1139/cjfas-2014-0473.

Chan, F. T., MacIsaac, H. J., & Bailey, S. A. (2016). Survival of ship biofouling assemblages during and after voyages to the Canadian Arctic. *Marine Biology, 163*(12), 14. https://doi.org/10.1007/s00227-016-3029-1.

Chavez, F. P., Ryan, J., Lluch-Cota, S. E., & Niquen, M. (2003). From anchovies to sardines and back: Multidecadal change in the Pacific Ocean. *Science, 299*(5604), 217–221. https://doi.org/10.1126/science.1075880.

Claudi, R., & Ravishankar, T. J. (2006). Quantification of risks of alien species introductions associated with ballast water discharge in the Gulf of St. Lawrence. *Biological Invasions, 8*(1), 25–44. https://doi.org/10.1007/s10530-005-0234-0.

Committee on Ships' Ballast Operations, N. R. C. (1996). *Stemming the tide: Controlling introductions of nonindigenous species by ships' ballast water*. Washington, DC: National Academy Press.

David, M., Perkovic, M., Suban, V., & Gollasch, S. (2012). A generic ballast water discharge assessment model as a decision supporting tool in ballast water management. *Decision Support Systems, 53*(1), 175–185. https://doi.org/10.1016/j.dss.2012.01.002.

Dawson, J., Johnston, M. E., & Stewart, E. J. (2014). Governance of Arctic expedition cruise ships in a time of rapid environmental and economic change. *Ocean & Coastal Management, 89*, 88–99. https://doi.org/10.1016/j.ocecoaman.2013.12.005.

DiBacco, C., Humphrey, D. B., Nasmith, L. E., & Levings, C. D. (2012). Ballast water transport of non-indigenous zooplankton to Canadian ports. *ICES Journal of Marine Science, 69*(3), 483–491. https://doi.org/10.1093/icesjms/fsr133.

Eguiluz, V. M., Fernandez-Gracia, J., Irigoien, X., & Duarte, C. M. (2016). A quantitative assessment of Arctic shipping in 2010–2014. *Scientific Reports, 6*. https://doi.org/10.1038/srep30682.

Eliasson, K., Ulfarsson, G. F., Valsson, T., & Gardarsson, S. M. (2017). Identification of development areas in a warming Arctic with respect to natural resources, transportation, protected areas, and geography. *Futures, 85*, 14–29. https://doi.org/10.1016/j.futures.2016.11.005.

Floerl, O., Rickard, G., Inglis, G., & Roulston, H. (2013). Predicted effects of climate change on potential sources of non-indigenous marine species. *Diversity and Distributions, 19*(3), 257–267. https://doi.org/10.1111/ddi.12048.

Gautier, D. L., Bird, K. J., Charpentier, R. R., Grantz, A., Houseknecht, D. W., Klett, T. R., Moore, T. E., Pitman, J. K., Schenk, C. J., Schuenemeyer, J. H., Sørensen, K., Tennyson, M. E., Valin, Z. C., & Wandrey, C. J. (2009). Assessment of undiscovered oil and gas in the Arctic. *Science, 324*(5931), 1175–1179. https://doi.org/10.1126/science.1169467.

Gollasch, S. (2002). The importance of ship hull fouling as a vector of species introductions into the North Sea. *Biofouling, 18*(2), 105–121. https://doi.org/10.1080/08927010290011361.

Gollasch, S. (2006). Overview on introduced aquatic species in European navigational and adjacent waters. *Helgoland Marine Research, 60*(2), 84–89. https://doi.org/10.1007/s10152-006-0022-y.

Gucu, A. C. (2002). Can overfishing be responsible for the successful establishment of Mnemiopsis leidyi in the Black Sea? *Estuarine Coastal and Shelf Science, 54*(3), 439–451. https://doi.org/10.1006/ecss.2000.0657.

H.R. 4283 — 104th Congress: National Invasive Species Act of 1996. November 21, 2016 https://www.govtrack.us/congress/bills/104/hr4283, (1996).

Hansen, C. O., Grønsedt, P., Lindstrøm Graversen, C., & Hendriksen, C. (2016). Arctic shipping – Commercial opportunities and challenges. *CBS Maritime*. ISBN 978-87-93262-03-4.

HELCOM. (2016). *Baltic marine environment protection commission – Helsinki Commission*. Retrieved from http://helcom.fi/about-us

Hewitt, C., & Campbell, M. (2010). The relative contribution of vectors to the introduction and translocation of invasive marine species. *Commonwealth of Australia*. ISBN 978-1-921575-14-3.

International Maritime Organization (IMO). (2004). International convention for the control and management of ships' ballast water and sediments, London, UK

IMO. (2007). Resolution MEPC.162(56). Guidelines for risk assessment under regulation a-4 of the BWM convention (G7).

IMO. (2008). Guidelines and guidance documents related to the implementation of the international convention for the control and management ships' ballast water and sediments, 2004. (Updated in 2015).

IMO. (2016). *International maritime organization*. Ballast Water Management. Retrieved from http://www.imo.org/en/OurWork/Environment/BallastWaterManagement/Pages/Default.aspx

IPCC. (2013). Climate change 2013: The physical science basis. In *Contribution of Working Group I to the fifth assessment report of the Intergovenmental Panel on Climate Change*. Cambridge/New York: Cambridge University Press.

Jackson, S. T., Betancourt, J. L., Booth, R. K., & Gray, S. T. (2009). Ecology and the ratchet of events: Climate variability, niche dimensions, and species distributions. *Proceedings of the National Academy of Sciences of the United States of America, 106*, 19685–19692. https://doi.org/10.1073/pnas.0901644106.

Jing, L., Chen, B., Zhang, B. Y., & Peng, H. X. (2012). A review of ballast water management practices and challenges in harsh and arctic environments. *Environmental Reviews, 20*(2), 83–108. https://doi.org/10.1139/a2012-002.

Lindegren, M., Checkley, D. M., Ohman, M. D., Koslow, J. A., & Goericke, R. (2016). Resilience and stability of a pelagic marine ecosystem. *Proceedings of the Royal Society B-Biological Sciences, 283*(1822), 9. https://doi.org/10.1098/rspb.2015.1931.

Mack, R. N., Simberloff, D., Lonsdale, W. M., Evans, H., Clout, M., & Bazzaz, F. A. (2000). Biotic invasions: Causes, epidemiology, global consequences, and control. *Ecological Applications, 10*(3), 689–710. https://doi.org/10.2307/2641039.

Maslowski, W., Kinney, J. C., Higgins, M., & Roberts, A. (2012). The future of Arctic Sea ice. In R. Jeanloz (Ed.), *Annual review of Earth and Planetary Sciences* (Vol. 40, pp. 625–654).

Miller, A. W., & Ruiz, G. M. (2014). Arctic shipping and marine invaders. *Nature Climate Change, 4*(6), 413–416.

Miller, A. W., Huber, T, Minton, M. S, & Ruiz, G. M. (2011). *Status and trends of ballast water management in the United States. Fourth Biennal report of the national ballast information clearinghouse*. Retrieved from United States Coast Guard (CG-5224) 2100 2nd Street, S.W. Washington, DC, 20593 USA.

Niimi, A. J. (2004). Environmental and economic factors can increase the risk of exotic species introductions to the arctic region through increased ballast water discharge. *Environmental Management, 33*(5), 712–718. https://doi.org/10.1007/s00267-004-3072-4.

Nonindigenous Aquatic Nuisance Prevention and Control Act of 1990 – Title I of P.L. 101–646 (104 Stat. 4761, 16 U.S.C. 4701, enacted November 29, 1990, (1990).

NSIDC. (2016). National Snow and ice data center. *Arctic Sea Ice News & Analysis*, September.

OSPAR. (2016). *OSPAR Commission protecting and conserving the marine environment of the North-East Atlantic*. Retrieved from http://www.ospar.org/about

OSPAR/HELCOM. (2015). Joint harmonised procedure for the contracting parties of OSPAR and HELCOM on the granting of exemptions under international convention for the control and management of ships' ballast water and sediments, Regulation A-4.

Philippart, C. J. M., Anadon, R., Danovaro, R., Dippner, J. W., Drinkwater, K. F., Hawkins, S. J., Oguz, T., O'Sullivan, G., & Reid, P. C. (2011). Impacts of climate change on European marine ecosystems: Observations, expectations and indicators. *Journal of Experimental Marine Biology and Ecology, 400*(1–2), 52–69. https://doi.org/10.1016/j.jembe.2011.02.023.

Pimentel, D., McNair, S., Janecka, J., Wightman, J., Simmonds, C., O'Connell, C.,. .. Tsomondo, T. (2001). Economic and environmental threats of alien plant, animal, and microbe invasions. *Agriculture Ecosystems & Environment,* 84(1), 1–20. doi:https://doi.org/10.1016/s0167-8809(00)00178-x.

Pizzolato, L., Howell, S. E. L., Dawson, J., Laliberte, F., & Copland, L. (2016). The influence of declining sea ice on shipping activity in the Canadian Arctic. *Geophysical Research Letters, 43*(23), 12146–12154. https://doi.org/10.1002/2016gl071489.

Rex, M. A., Stuart, C. T., & Coyne, G. (2000). Latitudinal gradients of species richness in the deep-sea benthos of the North Atlantic. *Proceedings of the National Academy of Sciences of the United States of America, 97*(8), 4082–4085. https://doi.org/10.1073/pnas.050589497.

Rhein, M., Rintoul, S. R., Aoki, S., Campos, E., Chambers, D., Feely, R. A., Gulev, S., Johnson, G. C., Josey, S. A., Kostianoy, A., Mauritzen, C., Roemmich, D., Talley, L. D., & Wang, F. (2013). Ocean. The physical science basis. Contribution of Working Group I to the fifth assessment report of the Intergovernmental Panel on Climate Change. In T. F. Stocker, D. Qin, G.-K. Plattner, M. Tignor, S. K. Allen, J. Boschung, A. Nauels, Y. Xia, V. Bex, & P. M. Midgley (Eds.), *Climate change.* Cambridge/New York: Cambridge University Press.

Stein, B. A., & Flack, S. R. (1996). *Americas least wanted: Alien species invasions of U.S. Ecosystem.* Arlington: The Nature Conservancy.

Sunday, J. M., Bates, A. E., & Dulvy, N. K. (2012). Thermal tolerance and the global redistribution of animals. *Nature Climate Change, 2*(9), 686–690. https://doi.org/10.1038/nclimate1539.

Tamelander, J., Riddering, L., Haag, F., & Matheickal, J. (2010). *Guidelines for development of National Ballast Water Management Strategies* (GloBallast monographs no. 18). Gland/London: GEF-UNDP-IMO GloBallast/IUCN.

Tilman, D. (1996). Biodiversity: Population versus ecosystem stability. *Ecology, 77*(2), 350–363. https://doi.org/10.2307/2265614.

Toph, A. (2013). Greenland opens up rare earth mining opportunities. *Rare Earth Investing News,* October 28.

Verna, D. E., & Harris, B. P. (2016). Review of ballast water management policy and associated implications for Alaska. *Marine Policy, 70,* 13–21. https://doi.org/10.1016/j.marpol.2016.04.024.

Walsh, J. E. (2013). MELTING ICE what is happening to Arctic Sea ice, and what does it mean for us? *Oceanography, 26*(2), 171–181.

Wang, M. Y., & Overland, J. E. (2009). A sea ice free summer Arctic within 30 years? *Geophysical Research Letters, 36.* https://doi.org/10.1029/2009gl037820.

Ware, C., Berge, J., Sundet, J. H., Kirkpatrick, J. B., Coutts, A. D. M., Jelmert, A., et al. (2014). Climate change, non-indigenous species and shipping: Assessing the risk of species introduction to a high-Arctic archipelago. *Diversity and Distributions, 20*(1), 10–19. https://doi.org/10.1111/ddi.12117.

Ware, C., Berge, J., Jelmert, A., Olsen, S. M., Pellissier, L., Wisz, M., Kriticos, D., Semenov, G., Kwasniewski, S., & Alsos, I. G. (2016). Biological introduction risks from shipping in a warming Arctic. *Journal of Applied Ecology, 53*(2), 340–349. https://doi.org/10.1111/1365-2664.12566.

Yasuhara, M., Hunt, G., van Dijken, G., Arrigo, K. R., Cronin, T. M., & Wollenburg, J. E. (2012). Patterns and controlling factors of species diversity in the Arctic Ocean. *Journal of Biogeography, 39*(11), 2081–2088. https://doi.org/10.1111/j.1365-2699.2012.02758.x.

Arctic Port Development

Julia Pahl and Brooks A. Kaiser

Abstract Melting Arctic sea ice, shore ice, and permafrost are changing costs and benefits to transport routes between Atlantic and Pacific oceans, and more generally, for maritime economic activity in the Arctic. We investigate the potential for development of Arctic ports from a logistics (demand) and an infrastructural (supply) point of view that directly incorporates local concerns. This approach broadens the scope of the discussion from existing analyses that focus primarily on the ways in which global forces, exerted through resource extraction or trans-polar shipping, impact the Arctic.

Keywords Arctic ports • Arctic marine infrastructure • Arctic economic development • Marine Arctic economy • Arctic Shipping

Introduction

Globalization and related commercial dynamism have grown through the increased ability to manage logistical supply chains, the increased capacity of port infrastructure, and evolution in means of transportation. These transformations of capacity and connectivity have enormous consequences on geography of places, commodity markets, and passenger flows (see Verny and Grigentin 2009). In this process, principal maritime commercial routes have to date changed very little compared to other aspects of globalization and trade. Global warming, however, is changing that. Three new Arctic maritime routes are becoming increasingly interesting to stakeholders ranging from shipping companies to local indigenous communities; see Fig. 1.

J. Pahl (✉)
SDU Engineering Operations Management, Department of Technology and Innovation, University of Southern Denmark, Campusvej 55, DK-5230 Odense, Denmark
e-mail: julp@iti.sdu.dk

B.A. Kaiser
Department of Sociology, Environmental and Business Economics, University of Southern Denmark, Niels Bohrs Vej 9, DK-6700 Esbjerg, Denmark
e-mail: baka@sam.sdu.dk

© Springer International Publishing AG 2018
N. Vestergaard et al. (eds.), *Arctic Marine Resource Governance and Development*,
Springer Polar Sciences, https://doi.org/10.1007/978-3-319-67365-3_8

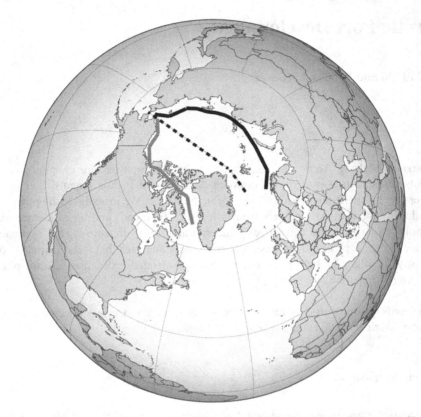

Fig. 1 Three Arctic routes: NWP grey line, NSR: black line, TPR: dotted line (Picture by: Chess-rat, CC BY-SA 3.0, http://creativecommons.org/licenses/by-sa/3.0), via Wikimedia Commons)

The Northern Sea Route (NSR; black line in Fig. 1) is becoming a viable option for containerized commodity flows between Asia and Europe, especially in light of current increasingly congested routes of maritime shipping such as the Suez Canal which serves the Asia-European market; see Verny and Grigentin (2009). The NSR is estimated to enable savings of 40% to 50% of sailing distance from Asia (Yokohama) to Europe (Rotterdam, (see Liu and Kronbak 2010; Stephenson et al. 2011), Fig. 2) and 40% less than the Suez Canal route (see Ircha and Higginbotham 2016). The Northwest Passage (NWP; grey line in Fig. 1) is less likely to be a significant route for cargo, with more dangerous navigation and less infrastructure, but its opening presents opportunities for *extractive* and *non-extractive* uses (see, e.g., Ircha and Higginbotham 2016), including tourism and the expansion of fisheries into new waters. Instead, the Trans-Polar Route (TPR; dotted line in Fig. 1) remains a hypothetical, but alluring future possibility for time- and money-saving trans-Arctic shipping; its fruition will change demand for the NSR through increased competition (see, e.g., Ircha and Higginbotham 2016).

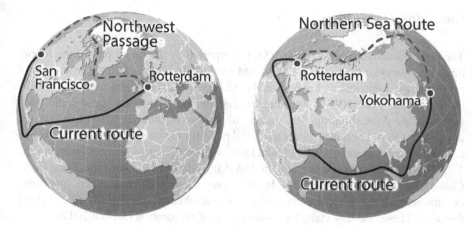

Fig. 2 NSR and NWP compared to current shipping routes (Picture credit: http://www.grida.no/graphicslib/detail/northern-sea-route-and-the-northwest-passage-compared-with-currently-used-shipping-routes_1336)

There is, however, a lack of Arctic marine infrastructure. This is slowing down the process for the regions to fully realize their resource potential, including the enhancement of lives for their population (see also Higginbotham and Grosu 2014; Ircha and Higginbotham 2016). Many Arctic communities depend on resupply via ocean or river transport. Such transport is efficient and relatively inexpensive. This dependence is increased by the dearth of reliable road connections. Such connections are expected to worsen for many communities in future with melting of the permafrost (see Higginbotham and Grosu 2014). Companies already active in Arctic shipping do express a great interest in extending their services (see Lasserre and Pelletier 2011), but with a focus on destination short sea shipping, motivated especially by servicing of mining (ore), processed metals, and oil and gas operations. This creates demand only at a limited scale in the short run, but might have high increases due to predicted LNG transportation from Russian Arctic gas fields to North America in the long run; see Lasserre and Pelletier (2011). Besides this, obstacles exist that stem from the relatively inhabited and geographically constrained characteristics of the traversed territories, e.g., along the Siberian coast between the Bering Strait and the Port of Murmansk. These characteristics do not permit stopovers, thoroughly reliable communication and navigation, and/or aid in case of emergencies (see Ho 2010; Verny and Grigentin 2009). These drawbacks continue despite increasing commitment to investment in search and rescue, particularly in Russian waters. Additional investments in infrastructure and marine services including safety and security are needed before the NSR can be subject to continuous and large-scale shipping.

Changing Demands

The potential attractiveness of the NSR is supported by the major part of the literature, e.g., Verny and Grigentin (2009) and Ho (2010), with significant reservations regarding the logistical and operational details (see Buixadé Farré et al. 2014). On the other hand, the NWP route remains remote, but appealing for specific uses beyond trans-polar shipping. The successful passage in summer 2016 of the luxury cruise ship Crystal Serenity serves as a focal point for the changing uses of the Arctic maritime environment. The ship, 250m long, with 68,870 gross tonnage, transported 1,750 passengers and crew through the NWP. It made stops at towns like Pond Inlet, Canada, population 1,549 (2011). In these very small ports, passengers and crew outnumbered local residents, taxing infrastructure heavily though for a very short duration. These one-day visits took over a year of preparation (Jepson 2017).

The TPR is likely eventually to eliminate most interest for trans-Arctic shippers in any port infrastructure within the Arctic, as it would enable rapid transit in open water unhindered by risky shorelines and national demands of the US, Canada or the Russian Federation. Today, this route remains hypothetical. Port investment decisions should still consider this approaching reality as one that limits long run port opportunities for transit trade but yet may increase demand for safety and security. With respect to shipping cargo, the three routes are substitutable competitors. But in other dimensions, particularly safety and security in the region, the three routes are more complements than substitutes. It is important to consider the economic, social, political and geophysical influences on the routes' futures jointly as well as separately.

Evolving Supply Provision

Port infrastructure decisions are increasingly under way in the Arctic today. Many plans for increased capacity began with increasing commodity prices at the beginning of the twenty-first century. Declining resource prices today, and disappointing returns on exploratory ventures in offshore drilling for oil, have curtailed interest, particularly in US, Canadian, and Danish (Greenlandic) waters. Very recent developments of onshore oil finds may be shifting interest back again (see discussion in Kaiser, Pahl and Horbel, Chap. 9 this volume). This volatility is part of the uncertainty that public planners must incorporate into decisions for port infrastructure investments and governance of port activities. Most new developments in the Western Arctic are thus focusing on other features for port infrastructure that can support local community needs. These include space for small local vessels and cruise tourism capacity. Icelandic and Norwegian decisions initiate from a much larger existing infrastructure base and populations, and focus on transforming port uses from coastal fisheries to a diversified portfolio of maritime activities. Meanwhile, in the Russian Arctic, more successful oil and gas ventures are initiating private investment in private ports.

Towards Arctic Port Development

This chapter builds on findings of the comprehensive book by Østreng et al. (2013) on shipping in Arctic waters. In that book, however, the authors take a primarily global view of the future and deduce impacts for the Arctic. Their approach is neither unusual nor ineffective for analyzing development influences on the Arctic. Global forces have long overwhelmed local and regional considerations in the sparsely populated area. We, however, move in the other direction, considering the local and regional perspective as the primary driving force in successful future Arctic infrastructure developments. Our Arctic-centric perspective aims to shift the conversation from imposed external governing and economic forces to one that considers options for more self-determination in future well-being.

In this chapter, we seek to identify how an Arctic port could serve to extenuate or eliminate several concerns pertaining to the passages themselves as well as the economic and environmental considerations that will drive their futures, and which physical and economic requirements such a port would need to fulfill. Moreover, we integrate the question of how port logistics demand and related required infrastructure have an influence on the development of Northern communities and their resources. We further discuss what actions are needed from a governance perspective in order to boost positive, and reduce negative, influences of economic development in the North.

To do this, we analyze the following aspects of demand and supply regarding the development of infrastructure, especially concerning the development of ports, that is required to assure safe and secure traffic in the Arctic. We hypothesize what would ideal Arctic ports look like to meet the needs of their multifaceted stakeholders, and where they should be further developed in order to spur meaningful diversified economic development. In general, in this chapter we provide support for policy that focuses port development more on the shore amenities and quality of life impacts to local and regional needs than on the port capacity for trans-Arctic shipping, or even destination shipping for resource extraction. The main rationale evolves from the combined hurdles presented to greater use of the NWP and NSR for trans-Arctic trade and the desire to improve quality of life that allows for preservation of cultural integrity for Arctic residents.

As stated in Ragner (2000), it is too early to think about shifting year-round Atlantic-Pacific trade routes to the Arctic, but not too early for concerned private and public interests to start planning. There is a high likelihood that ice covers will disappear and allow for longer seasonal, and eventually year-round, maritime transit operations. This melting simultaneously is expected to increase transportation costs by overland routes. This increases further the importance of port infrastructure in the region. We investigate the similarities and differences of the three route's needs and potentials in the context of port infrastructure investment to illustrate the scope of concerns for governance and coordination in Arctic development.

We sum up the impetus for development of ports in the Arctic in two contradictory thoughts. First, we cite McCague (2014) "if you build it, they will come.

If you don't, they will come anyway." On the other hand, the reverse statement for the Arctic may be equally accurate: without port infrastructure, the outside world may continue to bypass or exploit Arctic communities and thus separate their development paths from the rest of the world's gains from globalization and trade. The conflicting sentiments both lead to the same conclusion, which is that it is most advisable to be prepared.

State Analysis: Current and Future Strategies of Port Development

Which ports in the Arctic and surroundings have plans to expand according to projections of demand in the future? We take up this question in the following section. We first describe the current state of port infrastructure in the Arctic using the World Port Index (National Geospatial-Intelligence Agency 2016) and other publicly available information, including national Arctic strategic plans for building or extending ports. The most relevant ports with expansion plans are listed in Table 1 together with their infrastructure and inter-modal connections as well as plans for extensions according to future demand.

Table 2 summarizes port infrastructure more broadly in the Arctic. We note two key items. First, overall port capability in the Arctic is limited, with 135 ports identified overall. Almost 75% of them (99 ports) are classified as very small harbors and only one, Murmansk, is identified as a large harbor. The overall importance of these ports in their own comprehensive national port infrastructures also varies considerably across countries. For Greenland, all ports are considered Arctic ports; all policy decisions and investments will be made in this context. At the other end of the spectrum, the US only has 1.4% of its listed ports located in Arctic waters. It is unsurprising, then, that there is considerable variation in the interest in Arctic port infrastructure from these countries.

Present Arctic Strategies

The 2009 Arctic Marine shipping assessment report (see Arctic Council 2009) highlights three broad and inter-related themes fundamental to the understanding and evaluating Arctic marine shipping:

1. Enhancing Arctic marine safety,
2. Protecting Arctic people and their environment, and
3. Building Arctic marine infrastructure.

The study recommends developing improved navigational aids, e.g., ship routing, vessel tracking, traffic separation, and identification of areas of special concern (see

Table 1 Current and planned relevant (deep-water) ports in the Arctic

Port	Location	Country code	Population	Regional per capita income	(max) Water depth (in m) at berth	Current use			Current connectivity			Plans/comments
						Cargo (B./C.)[a]	Fishing	Cruise	Land	Air	All year	
Port of Nome	Western Alaska, Bering sea	US	3,598 (2010)	$23,402	6.9	B	Yes	Yes	No	Yes	No	Deep-water port extension to 8.5 m in 2020 planned
Port of Iqaluit (planned)	Nunavut	CA	6,699 (2011)	$41,100 (Nunavut)	11	B/C	Yes	Yes	Few	Yes	all year	Deep-water port to be opened in/around 2020 planned
Port of Churchill	Prairie region, Hudson Bay, Northern Manitoba	CA	813 (2011)	$33,650 (Manit.)	10	B	No	Yes	Yes	No	No	Only deep-water ocean port in Prairie region; accommodation of Postpanamax ships; closed in 2016, 90% grain transport, petroleum
Port of Nanisivik	Arctic Bay, Nunavut, near Baffin Bay	CA	0	$41,100 (Nunavut)	n.a. (not available)	B/C?	No	No	No	No	No	Naval refueling facility for Navy ships planned to open in 2018
Port of Finnafjördur (planned)	North-East of Iceland	IS	1,300 (2010)	$41,160 (2013, Iceland)	n.a. (not available)	B/C?		No	Yes	Yes	Yes	Planned for Oil and gas operations, as hub for trans-Arctic shipping, and offshore Petroleum, LNG bunkering, and SAR facility

[a] B bulk (OG oil and gas), C container

Table 2 Arctic (national) ports by size

Country	Large	Medium	Small	Very small	Total ports	Percent of ports in the Arctic
Canada	0 (4)	1 (14)	3 (72)	14 (196)	18 (286)	6.3%
Greenland	0 (0)	0 (0)	8 (8)	13 (16)	21 (24)	90.0%
Iceland	0 (0)	0 (2)	0 (2)	13 (24)	13 (28)	46.4%
Norway	0 (1)	3 (10)	9 (34)	35 (90)	47 (135)	34.8%
Russian Federation	1 (4)	2 (5)	9 (21)	12 (44)	24 (74)	32.4%
United States	0 (21)	0 (38)	0 (132)	9 (475)	9 (666)	1.4%
International (Svalbard)	0	0	0	3	3	
Total Arctic ports	**1**	**6**	**29**	**96**	**135**	

U.S. Army Corps of Engineers 2013). There are joint efforts established within the Barents Euro-Arctic Council's Joint Barents Transport Plan that include Norway, Sweden, Finland, and Russia (Higginbotham and Grosu 2014). Similar efforts should be encouraged in North America and the North Atlantic.

Arctic ports need to transform to meet the needs of their constituents. The literature to date regarding Arctic port investment generally focuses more heavily on the interests and needs of shippers and resource extractors than it does on local communities. When it does include discussions of indigenous populations, the issues still focus primarily on North-South connections that support resource extractive trade (see, e.g., Østreng et al. 2013). This is a reasonable by-product of the high measurable value of this potential in, e.g., days saved at sea or oil extracted. These discussions are not, however, comprehensive assessments of the value changes that will stem from changes in accessibility and port infrastructure in the Arctic. We consider the supply of port structure more specifically from the local and regional perspective to gain insights into the prospects and perils involved. This focus stems not only from the lack of existing attention to the local and regional perspective but also due to our assessment that the shore-based routes are unlikely to serve, economically or technically, as consistent high-volume cargo shipping pathways for very many years. This is because the current climate change trajectory (or its intensification) will open the TPR so that the route becomes the economically preferred option.

From an infrastructural point of view, it is interesting to investigate if there would be demand from shipping lines for Arctic ports and what shipping lines would require in terms of services and infrastructure to call at Arctic ports. How would these requirements regarding the infrastructure and offered services of a port influence the development of Northern communities? What are the infrastructural costs of port development, taking into account permafrost issues and melting ice due to global warming?

Figure 3 maps existing or planned Arctic deep-water ports as deep drafts are required to accommodate, e.g., fully loaded Panamax ships. A deep-water port can be defined as one that can accommodate large heavy loaded ships requiring at least

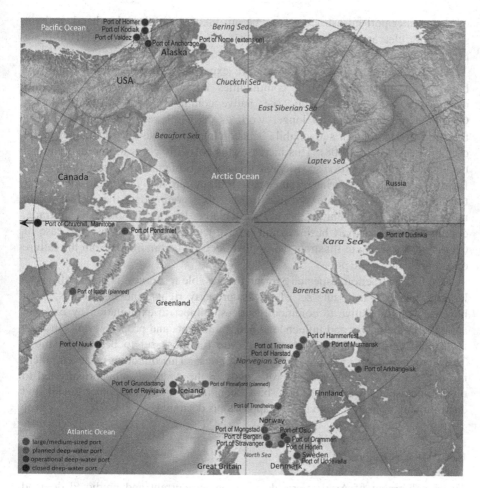

Fig. 3 Operating and planned medium-sized and/or large-sized (deep-water) ports in the Arctic Ocean

a 30 feet water depth or more, which is equivalent to approximately 9.1 m.[1] Other sources state that a draft of at least 12.04 m is needed.[2] While the relatively small Panamax ship size might at first glance appear interesting for operations in the Arctic for bulk and container shipping, recent synthetic research from the Copenhagen Business School on the viability of Arctic Shipping supports our contention that such shipping will not occur at high levels (Hansen et al. 2016). Thus the focus

---

[1]See http://www.marineinsight.com/ports/what-are-deep-water-ports/, last call: 18 Dec 2016 as well as Dasgupta (2016).

[2]See http://www.marineinsight.com/types-of-ships/the-ultimate-guide-to-ship-sizes/, last call: 18 Dec 2016.

on deep water port infrastructure may be misdirected. On the other hand, without such depth clearance then the projections of low use will become self-fulfilling prophesies.

Almost regardless of depths, ports are expected to become more important hubs within and amongst Arctic communities. Existing technologies and the use of transport in the Inuit Arctic (i.e. native communities in Northern Alaska, Canada, Greenland) can no longer work as efficiently as they once did, e.g., transport by sled dog or snowmobile is becoming unreliable due to changes in ice quality (Pearce et al. 2008). This increases the costs of access to marine mammal resources and reduces food security and other aspects of cultural heritage and value. Pathways forward for marine resource based cultures are likely to require significant transitions in port structure and use in order to accommodate these changes. In more developed communities already participating in considerable external trade, continued health of port business and infrastructure is vital. We use the case of Churchill, Manitoba to illustrate the ripple effects on communities from the failure of a significant Arctic port.

Resource extraction, especially for non-renewable resources, is potentially very lucrative under a favorable set of prices, property rights, and institutions. It is, however, very sensitive to commodity prices, subject to exploitation by external actors with the power and influence to determine rents and property rights in their favor, and currently not very attractive even for those set to benefit the most. This is mainly due to volatile and declining prices. This direction of development through resource extraction, tried repeatedly in the Arctic, has historically led to disenfranchisement, cyclical boom-bust economic activity and other social and environmental ills. A more balanced approach is desirable.

U.S. (Alaska) Region

The development of Arctic ports depends on the current and predicted demand situation. Demand declines or non-realization of predicted demands have direct influence of port development. For instance, in 2013, there were deep-draft ports in Anchorage, Seward, Valdez, Kodiak, Unalaska (Dutch Harbor), and Homer, but none along Alaska's Arctic coastline (see U.S. Army Corps of Engineers 2013). Table 3 lists vessel trips of ships greater than 100 feet that traversed the Bering Strait into (or from) Arctic waters from five different Alaskan areas. The North Slope and the Bering Strait can be considered Arctic, Nome (Norton Sound) is the entrance area to the Arctic, and the Pribilofs and Southwest Alaska are sub-Arctic. Fishing vessels are excluded. The data show traffic through the Bering Strait uses the Nome area most, though there is high variability from year to year. Northern and southern traffic do not show much overlap, though trips from sub-Arctic Alaskan waters have increased tenfold (from 19 to 192) in three year period for which we have data. North Slope vessel traffic is limited and was falling for the period in question.

An updated study with 2013–2015 data visually examines all types of vessel traffic through the Bering Strait, see Fig. 4. In the figure, one sees that bulk cargo

Table 3 Total vessel trips, by region and year (Source: U.S. Army Corps of Engineers 2013)

Region	2009	2010	2011
Bering Strait, NW Alaska	191	286	255
Nome, Norton Sound	379	675	402
North Slope	32	25	21
Pribilofs	16	86	103
Southwest Alaska	3	74	89

Fig. 4 AIS tracks by vessel type in the bering straight in between 2013–2015 (Source: Fletcher et al. 2016)

through the strait almost exclusively services the Red Dog Mine (zinc and lead) on the US side (with 92% of vessel operating days). All other types of vessels use a much greater diversity of ports. From this we conclude that increased bulk cargo from mining enterprises has a lower likelihood of generating economies of scope that extend to other community economic activities than other utilization.

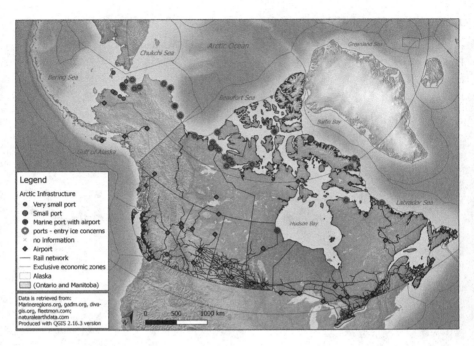

Fig. 5 Canadian and Alaskan ports

Figure 5 provides an overview on all currently operating Canadian and Alaskan ports that are listed in the World Port Index (see National Geospatial-Intelligence Agency 2016). Table 4 provides more details on the ports. The entries are sorted by harbor size from large sizes descending to small sizes. As depicted in the table, most of the Arctic ports in the U.S. and Alaska are classified as very small, and have limited cranes and repair facilities.

A 2013 U.S. Army Corps of Engineers study (ACE) (U.S. Army Corps of Engineers 2013) and a 2016 Ocean Conservancy report (Fletcher et al. 2016) give guidance on deep-draft infrastructure development for federal, state, local, and private investors in order to respond to the increased needs regarding Arctic traffic and resources. These studies explicitly include expected increases in risks of incidents; the Ocean Conservancy report focuses on oil exposure risks. The U.S. Coast Guard's ability to respond is an essential component; however the current US administration's desire to significantly reduce the U.S. Coast Guard's budget threatens safety and security in the region. Indeed, the uncertainty over the government budget and the severity of the proposed cuts (\$ 1.3 billion, or 12% of the USCG budget (U.S. White House 2017b)) jeopardizes actual USCG presence in the Arctic altogether. Furthermore, inadequate Coast Guard support hinders private investment interest due to the inability to guarantee safety and security (Gramer 2017; Østhagen 2015).

Table 4 Details on current and planned ports in the US and Alaska

Port	Population (2010)	Harbor size	Harbor type	Water depth (in m)ᵃ	Shelter	Tide range	ETA message	Pilot available	Fixed crane	Crane mobile	Crane float	Provisions	Water	Fuel/oil	Diesel	Deck supply	Repair code	Dry dock	Marine railway	Plans/comments
Nome	3,598	S	OR	6.9	P	2	Y	Y				Y	Y	Y	Y	Y				Extension to 8.5 m by 2020
Kwiguk		V	RN	n.i.	F	1	Y	Y				Y	Y	Y						
Kotlik		V	RN	n.i.	F	3	Y	Y				Y	Y	Y	Y					
Hamilton		V	RN	n.i.	F	3	Y	Y						Y	Y					
St. Michael		V	OR	n.i.	F	4	Y	Y				Y			Y		C			
Unalakleet		V	OR	n.i.	P	3	Y	Y				Y								
Solomon		V	OR	n.i.	P	1	Y	Y												
Grantley Harbor	229	V	OR	n.i.	G	1	Y	Y				Y							S	
Savoonga		V	OR	n.i.	P	1	Y	Y				Y	N	Y	Y	N				
Shishmaref	563	V	CN	n.i.	F	0		Y				Y	Y	Y	Y					
Deering	122	V	OR	n.i.	F	2		Y				N	N	N	N	N				
Kotzebue	3,201	V	CN	n.i.	F	2		Y												
Wainwright (Ulguniq)	556	V	OR	n.i.	P	1		Y												
Barrow (Utqiagvik)	4,212	V	OR	n.i.	P	0		Y	Y	Y	Y	N								
Prudhoe Bay	2,174	V	CN	n.i.	P	0				Y	N	Y								
Bernard Harbor	239	V	CN	n.i.	P	0														

Harbor Size: *L* large, *M* medium, *S* small, *V* very small, Columns containing Y or N: *Y* yes, *N* no
Harbor Type: *OR* open roadstead, *CN* coastal natural, *RN* river natural, Shelter: *E* excellent, *G* good, *F* fair, *P* poor, *N* none
Water depth: *n.i.* not indicated
Repair Code: *A* major, *B* moderate, *C* limited, *D* emergency only, *N* none, Marine Railway: *L* large, *M* medium, *S* small
ᵃWater depth (in m) is max. at berth

The ACE study reflects not only on economic parameters, but also on rural communities working to maintain their subsistence lifestyles. This is necessary as their marine mammal food resources will be affected by increased traffic. The analysis takes into account the marine infrastructure from Bethel west up along the north to the east until the Canadian border. Interesting candidate sites according to evaluation criteria that include port proximity to mining/oil/gas, inter-modal connections, natural water depth, and navigation accessibility have been evaluated. From this evaluation, the US ACE shortlisted the Port of Nome, Port Clarence, Cape Darby, and Barrow (now Utqiagvik) for further development. Shell Oil's 2015 retreat from the Arctic, sparked by declining oil prices and failure to make a major oil find at their offshore Burger J well, was interpreted as a death knell for Alaskan oil exploration. Conditions continue to change rapidly in the Arctic, however. Major recent finds by Caelus Energy (Wald 2016) and Repsol (Jenkins 2017) are rapidly reigniting interest. The Chukchi, Beaufort and Bering Seas are identified as having high levels of oil resources. Renewed offshore interest, and increasingly greater amounts of it, if oil and gas prices increase, is expected. The first of these renewed exploration activities, proposed by Eni Corp. (with cooperation from Shell and Repsol) for the Beaufort Sea is currently under consideration by the Bureau of Ocean Energy Management (BOEM) under the new, more resource-extraction oriented U.S. federal administration (Dlouhy 2017).

In 2015, the port of Nome was selected as the preferred site for a deep-draft port to handle ocean-going ships (see also Ircha and Higginbotham 2016). The port of Nome is a regional transshipment hub for western Alaska communities for items such as heating oil, gasoline, construction supplies, non-perishable food items, gravel, etc. (see McDowell Group 2016). Nevertheless, efforts have stopped based on the decision of Shell to end drilling in U.S. Arctic waters (see DeMarban 2015). This does not square well with the broader community needs for developing the Arctic infrastructure that was recognized by the Obama administration. It highlights the region's current dependence on volatile global commodity markets and the need to reduce this dependence. Signals from the current administration favor oil exploration (U.S. White House 2017c) and infrastructure investment (U.S. White House 2017a), but there are simultaneous proposals to significantly reduce funding for supporting agencies, including the U.S. Coast Guard and the National Oceanic and Atmospheric Administration (U.S. White House 2017b). The end effect on Nome's investment decision remains uncertain. The city of Nome has a broader interest in expanding port activities to go beyond the economic benefits of oil and gas industries (see DeMarban 2015) as they are experiencing growing vessel traffic. The interest of Nome city is especially geared toward local communities having more influence over the general development of port infrastructure in the region. This is supported by a great amount of the Arctic port development literature as well as in the daily news. The latter also emphasizes the need for a coordinated global effort for the development of Arctic port infrastructure not only for economic reasons, but also for highly vulnerable marine ecosystems and indigenous communities that depend upon them (see Chambers 2015).

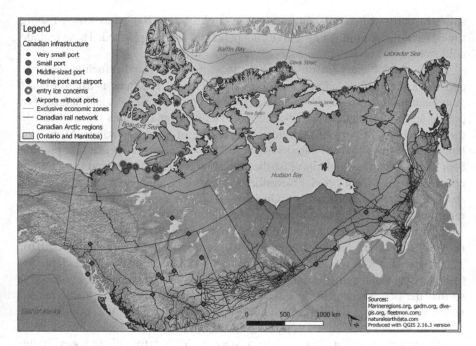

Fig. 6 Canadian ports

Canadian Region

Canadian Arctic shipping includes both destination and transit shipping with the first being the most frequent type (see Ircha and Higginbotham 2016). Destination shipping further encompasses mainly resupply (all types of products including consumer goods, etc.), oil and gas (mineral) resources transport, fishing, government research, cruise ships, as well as Canadian Coast Guard ice-breaking activities (see Ircha and Higginbotham 2016). Transit shipping passes from the Atlantic to/from the Pacific (see Ircha and Higginbotham 2016) (Fig. 6).

A port at Iqaluit is discussed in Aarluk Consulting et al. (2005), though it is not indexed in the World Port Index (see National Geospatial-Intelligence Agency 2016 and Frizzel 2017). Plans to extend this port to a deep-water port are in place (see Aarluk Consulting et al. 2005) with the port opening in/around 2020 (see, e.g., Ducharme 2016; Van Dusen 2016). The plans encompass the development of sufficient berthing and facilities for docking and cargo handling during the day and a secure holding area for cargo (see Aarluk Consulting et al. 2005). Petroleum is handled as well, but the current petroleum handling suffers from proper equipment and induces high environmental risks. A hydraulic oil transfer arm with direct access to the onshore petroleum pipeline is planned.

Regarding fisheries, high fishing quotas for shrimps and turbot have been allocated to Nunavut organizations, but fishing boats are forced to depart from southern ports with significant transit times to Frobisher Bay (thus losing fishing time). Facilities for unloading and storing palletized fish are planned. These include cold storage and reefer vessel access.

Cruise ships bypass Iqaluit due to the lack of resupply and refueling facilities and due to the lack of landing facilities for passengers. The same is true for coast guard and military vessels. A high tide range of 12 m adds to risks and damages to people and equipment. The deep-water port facilities can ease this risk and enable the accessibility of port infrastructure independent of the tides' waters. Moreover, landing facilities for small cargo vessels and also local outfitters that can provide boat tours to visitors would increase the utilization level of the port and increase job opportunities for local people. Nevertheless, more consultation is needed to create space for local fishery activities. Questions from the local community also arise regarding roads to and from the port (see Ducharme 2016).

Infrastructure for resupplies for northern villages is lacking, so that mostly, resupply ships carry landing barges on board (see Ircha and Higginbotham 2016). This is also due to high tide ranges for many Canadian Arctic ports (see Table 5). Cumbersome and lengthy operations for landing cargo and passengers are necessary, so that the development of deep-water ports in such areas with high tide ranges would be a possible solution in relation to perspective demand viz. utilization of such port facilities.

The current situation has lead to a stop in drilling for oil and gas in the Canadian Arctic since 2006. Reasons given by firms for the stopping are: inadequate infrastructure for transport, including lack of pipelines and roads; poor oil-spill response capabilities; and regulatory and environmental burdens (see Ircha and Higginbotham 2016 and the references therein) that have been brought to light in the case of Iqaluit (see Aarluk Consulting et al. 2005). The Canadian government is working on the infrastructure problem by working to connect Inuvik and Tuktoyaktur overland (see Ircha and Higginbotham 2016), by a road scheduled to open in 2017 (see Barton 2016).

Icelandic Region

Iceland has extensive plans for port development. For instance, Icelandic plans include evolving beyond fishing activities which have long been the major marine elements of Iceland. Instead, the nation wants to achieve a prominent role in maritime logistics (see Bremenports 2013). An example is the Finnafjordur project that includes the planning of a new port in the north-east of Iceland with year-round ice-free waters due to the Gulf stream. German expertise has been sought to develop the plans (see Bremenports 2013). Besides the Finnafjordur project, a deep-water port is contemplated at Isafjordur, according to Ircha and Higginbotham (2016). It should serve as a transshipment hub for containers coming from the Arctic to eastern North America and Europe. This activity currently is performed by Russia's Murmansk port (see Ircha and Higginbotham 2016). Isafjordur now mainly serves as a port for cruise ships and fishing activity.

Table 5 Details on current and planned ports in Canada

Port	Population (2010)	Harbor size	Harbor type	Water depth (in m)[a]	Shelter	Tide range	ETA message	Pilot available	Fixed crane	Crane mobile	Crane float	Provisions	Water	Fuel/oil	Diesel	Deck supply	Repair code	Dry dock	Marine railway	Plans/comments
Pond Inlet	1,549	M	OR	n.i.	P	6														
Paulatuk	313	S	CN	1.2	F	2							Y							
Churchill	813	S	RN	11.5	F	5	Y	Y		Y		Y	Y	Y	Y		B		L	Deep-water port, closed in 2016
Bernard Harbor		V	CN	11	P	0														
Tuktoyaktuk	854	V	CN	4–6	F	1	Y	Y				Y	Y	Y	Y		A	M		Hub for oil and gas
Police Point		V	CN	n.i.	F	2						N		N	N					
Pearce Point		V	CN	n.i.	G	1														
Tysoe Point		V	CN	n.i.	P	1														
Sachs Harbor	112	V	CN	1.8	P	2							Y							
Cape Young		V	OR	n.i.	P	0														
Lady Franklin Port		V	CN	n.i.	F	0							Y							
Coppermine (Kugluktuk)	1,459	V	RN	n.i.	F	1								Y			C			
Cambridge Bay	1,477	V	CN	3.2	G	1						Y	Y	Y			C			Airport in vicinity

(continued)

Table 5 (continued)

Port	Population (2010)	Harbor size	Harbor type	Water depth (in m)[a]	Shelter	Tide range	ETA message	Pilot available	Fixed crane	Crane mobile	Crane float	Provisions	Water	Fuel/oil	Diesel	Deck supply	Repair code	Dry dock	Marine railway	Plans/comments
Resolute Bay	229	V	CN	8.5	F	6				Y			Y							Airport in vicinity
Padloping Island		V	CN	n.i.	F	0														
Pangnirtung	1,425	V	CN	n.i.	F	15			Y	Y										
Saglek Bay		V	CN	n.i.	F	3	N													
Kangiqsujuaq (Maricourt)	479	V	CN	n.i.	F	5	N	N					Y							
Bathurst Inlet				n.i.																Planned private port for mining companies; not listed in the World Port Index
Nanisivik				10																Refueling facility for Naval ships; airport in vicinity
Iqaluit	6,699			11																Planned deep-water port; high tidal range, airport in vicinity

Harbor Size: *L* large, *M* medium, *S* small, *V* very small; Columns containing *Y* or *N*: *Y* yes, *N* no
Harbor Type: *OR* open roadstead, *CN* coastal natural, *RN* river natural; Shelter: *E* excellent, *G* good, *F* fair, *P* poor, *N* none
Water depth: *n.i.* not indicated
Repair Code: *A* major, *B* moderate, *C* limited, *D* emergency only, *N* none; Marine Railway: *L* large, *M* medium, *S* small
[a]Water depth (in m) is max. at berth

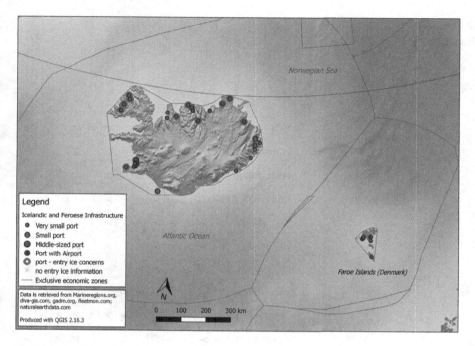

Fig. 7 Icelandic ports

Iceland predicts major economic potential with respect to oil and natural gas deposits, particularly off the Northeast coast, though exploration remains in early phases (National Energy Authority of Iceland (Orkusofnun) 2017). Moreover, the increasing cruise shipping business in Arctic waters is evaluated by Icelandic authorities as another important factor that necessitates new port infrastructure. This is also reflected in Table 6, showing that Iceland has 25 very small ports, one small sized port, and two medium sized ports identified; see also Fig. 7 that presents all existing ports of Iceland.

Norwegian Region

Norway is one of the best prepared Arctic countries with respect to port infrastructure. Its largest port is the port of Oslo with good infrastructure including efficient cargo terminals and short distance connections to railway and road. Though not in the Arctic, it serves as a gateway for cruise and ferry ships to the region. Additionally, Norway also has three medium sized ports in the Arctic and another seven to the south that are well oriented toward Arctic use and to serve for ships on the NSR. A selection of the ports are listed in Table 7, while Fig. 8 illustrates all Norwegian ports north of the Arctic circle in addition to the main southerly ports. Public plans for creating new ports or extending existing ports currently do not exist.

Table 6 Details on current and planned ports in Iceland[a]

Port	Population (2010)	Harbor size	Harbor type	Water depth (in m)[a]	Shelter	Tide range	ETA message	Pilot available	Fixed crane	Crane mobile	Crane float	Provisions	Water	Fuel/oil	Diesel	Deck supply	Repair code	Dry dock	Marine railway	Plans/comments
Reykjavik		M	CB	n.i.	F	4	Y	Y		Y	Y	Y	Y	Y	Y	Y	B	L	L	
Grundartangi		M	CB	n.i.	G	5	N	Y		Y		Y	Y	Y	Y		C			
Seydhisfjordhur		S	CN	n.i.	F	2	N	Y	Y	Y		Y	Y	Y			B		S	
Keflavik	15,129	V	OR	12.5	P	5	Y	Y		Y		Y	Y	Y	Y	Y	C		S	
Hafnarfjordur		V	CN	n.i.	F	4	Y	Y		Y		Y	Y	Y	Y	Y	C		M	
Bildudalur	166	V	CN	n.i.	F	3					Y	Y	N	N	N		C			
Blonduos		V	OR	n.i.	P	3					Y	Y								
Siglufjordhur	1,206	V	CN	n.i.	F	2	Y	Y	N	N	N	Y	Y	Y	Y	Y	B		S	
Olafsfjordhur	824	V	CN	n.i.	P	2	N												S	
Akureyri	18,191	V	CN	n.i.	F	2	Y	Y	Y	N	N	Y	Y	Y	Y		C	M	S	
Kopasker	120	V	CN	n.i.	P	1	N	Y	N	N	N					N				
Vopnafjordhur	668	V	CN	n.i.	F	2		Y			Y	Y	Y	Y						
Eskifjordhur	1,043	V	CN	n.i.	F	2	N	Y		Y	Y	Y	Y	Y			C		S	

Port	Water depth (m)ᵃ	Harbor Size	Harbor Type	Water depth	Shelter									Repair Code	Marine Railway	Notes
Hornabjordur		V	CN	n.i.	F	4	Y		Y			Y	Y		S	
Straumsvik		V	CN	n.i.	F	0	N	Y								
Bolungavik	882	V	CB	n.i.	F	0	N	Y								
Dalvik		V	CB	n.i.	F	0										
Skerjafjordur		V	CN	n.i.	G	4	Y			Y	Y	Y	Y	C		
Thingeyri	260	V	CN	n.i.	E	3			Y	Y	N	N	N	C		
Isafjordur	2,600	V	CN	n.i.	G	2	Y	Y	Y	Y	Y	Y	Y	B	M	Cruise ship port; considered to be extended to a deep-water port
Skagastrond		V	CN	n.i.	G	2	N				Y	Y	C		S	
Saudarkrokur	2,572	V	CN	n.i.	P	1	Y		Y	N	N	N		C		
Husavik		V	CN	n.i.	P	1	N		Y	Y	Y					
Raufarhofn		V	CN	n.i.	P	2	Y		Y	Y	Y	Y	Y	B		
Neskaupstadur		V	CN	n.i.	F	2	Y		Y	Y	Y	Y	C		S	
Budir		V	CN	n.i.	F	2	N		Y	Y	Y	C			S	
Djupivogur		V	CN	n.i.	F	1	Y			N	N	N	N			
Vestmannaeyjar		V	CN	n.i.	E	3	Y	Y	Y	Y	Y	Y	C		S	

Harbor Size: *L* large, *M* medium, *S* small, *V* very small, Columns containing Y or N: *Y* yes, *N* no

Harbor Type: *OR* open roadstead, *CN* coastal natural, *RN* river natural, Shelter: *E* excellent, *G* good, *F* fair, *P* poor, *N* none

Water depth: *n.i.* not indicated

Repair Code: *A* major, *B* moderate, *C* limited, *D* emergency only, *N* none, Marine Railway: *L* large, *M* medium, *S* small

ᵃWater depth (in m) is max. at berth

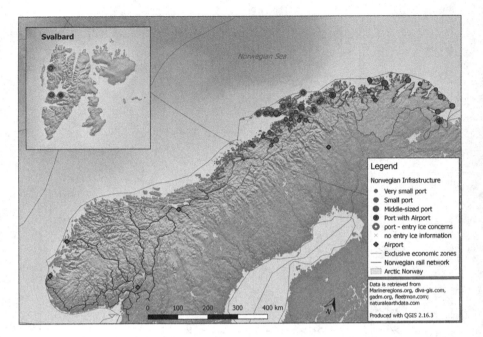

Fig. 8 Norwegian ports

Russian Federation Region

Russia is the primary stakeholder when it comes to the NSR due to its position. The Russian Federation encompasses almost all of the NSR; there are 24 existing ports, see Fig. 9. The largest is the sea port of Murmansk, which has a natural river entrance. Ports with medium size are Dudinka and Arkhangels'k that are also accessible via a river (see National Geospatial-Intelligence Agency 2016). Moreover, Russia has 20 small or very small ports on the shores of the NSR. The population of the Russian Arctic, and the population living along the NSR, is considerably greater than along the NWP. Most Norwegian and Russian communities are also more market- and/or military- oriented and globally interdependent in daily life than the communities along the NWP. These ports service naval operations and trade, and several have done so for decades. Improved marine transportation thus will affect many more lives directly (Table 8).

The Russian government has released a development plan for the NSR from now to 2030 that highlights the need to provide support for safer and more reliable navigation for ships on the NSR that are exporting Russian natural resources, as well as for those transiting with international cargo transport (see also Gunnarsson 2015). The main cargo types transported in 2015 on the Russian parts of the NSR were goods and project cargo. Project cargo includes materials to build the future Yamal LNG plant at the port of Sabetta on the Yamal peninsula, further increasing

Table 7 Details on current and planned ports in Norway (selection)

Port	Population (2010)	Harbor size	Harbor type	Water depth (in m)ᵃ	Shelter	Tide range	ETA message	Pilot available	Fixed crane	Crane mobile	Crane float	Provisions	Water	Fuel/oil	Diesel	Deck supply	Repair code	Dry dock	Marine railway	Plans/comments
Oslo		L	CN	11	G	1	Y	Y	Y	Y	Y	Y	Y	Y	Y	Y	A		L	
Hammerfest	7,568	M	CN	n.i.	G	3	Y	Y	Y	Y		Y	Y	Y	Y		C			
Tromso	58,486	M	OR	n.i.	G	2	Y	Y	Y	Y		Y	Y	Y	Y		B	M	M	
Harstad	23,640	M	CN	n.i.	G	2	Y	Y		Y	Y		Y	Y	Y	Y	A	L	L	
Trondheim		M	CN	n.i.	E	2	Y	Y	Y	Y	Y	Y	Y	Y	Y		C	M	L	
Mongstad		M	OR	n.i.	G	7	Y	Y	Y	Y	Y	Y	Y	Y	Y	Y	C	M		
Bergen		M	CN	n.i.	E	1	Y	Y	Y	Y		Y	Y	Y	Y	Y	A	M	S	
Kastro		M	OR	n.i.	F	n.i.	Y	Y	Y		Y	Y	N	N			C	M		
Stravanger		M	CB	n.i.	E	1	Y	Y	Y	Y	Y	Y	Y	Y	Y	Y	B	L		
Horten		M	CN	n.i.	G	1	Y	Y		Y	Y		Y	Y		Y	A	M	M	
Drammen		M	CB	n.i.	G	n.i.	Y	Y		Y	Y	Y	Y	Y	Y		B		L	
Mo Inlet		S	OR	n.i.	E	3	Y	Y	Y	Y			Y	Y	Y	Y	C		M	
Svolvaer	4,487	S	CN	n.i.	G	3	Y	Y		Y	Y		Y	Y	Y	Y	C	M		

(continued)

Table 7 (continued)

Port	Population (2010)	Harbor size	Harbor type	Water depth (in m)[a]	Shelter	Tide range	ETA message	Pilot available	Fixed crane	Crane mobile	Crane float	Provisions	Water	Fuel/oil	Diesel	Deck supply	Repair code	Dry dock	Marine railway	Plans/comments
Lakselv	2,258	S	CN	n.i.	G	0	Y	N	N				Y	N	N	N	N			
Kirkenes	3,498	S	CN	n.i.	G	3	Y	Y	Y	Y			Y	Y	N	N	N			
Vadso	5,116	S	CB	n.i.	F	3	Y	Y					Y	Y	Y	N	C		S	
Batsfjorden	2,133	S	CN	n.i.	G	0	Y	Y	Y	Y			Y	Y	Y	Y	C			
Vardo	1,893	S	CN	n.i.	E	3	Y	Y		Y	N		Y	Y			C			
Honningsvag	2,415	S	CN	n.i.	F	0	Y	Y	Y	Y			Y	Y	Y		C			
Sorvaer	201	V	CN	n.i.	P	0	Y		Y			Y	Y		Y					
Bredvik		V	CN	n.i.	G	0	Y	Y	N	N	N	Y	Y		Y	N	N		S	
Oksfjord	506	V	CN	n.i.	G	0	Y					Y	Y	Y						
Alta	20,000	V	CN	n.i.	G	3	Y					Y	Y		Y					

Harbor Size: *L* large, *M* medium, *S* small, *V* very small, Columns containing Y or N: *Y* yes, *N* no

Harbor Type: *OR* open roadstead, *CN* coastal natural, *RN* river natural, Shelter: *E* excellent, *G* good, *F* fair, *P* poor, *N* none

Water depth: *n.i.* not indicated

Repair Code: *A* major, *B* moderate, *C* limited, *D* emergency only, *N* none, Marine Railway: *L* large, *N* none, *M* medium, *S* small

[a]Water depth (in m) is max. at berth

Table 8 Details on current and planned ports in Russia

Port	Population (2010)	Harbor size	Harbor type	Water depth (in m)[a]	Shelter	Tide range	ETA message	Pilot available	Fixed crane	Crane mobile	Crane float	Provisions	Water	Fuel/oil	Diesel	Deck supply	Repair code	Dry dock	Marine railway	Plans/comments
Murmansk	299,148	L	RN	10	G	3	Y	Y	Y	Y	Y		Y	Y	Y	Y	A	L	M	
Dudinka	22,175	M	RN	11.8	G	0				Y	Y			Y	Y		C			
Arkhangels'k	348,783	M	RN	9.2	G	1	Y	Y	Y	Y	Y	Y	Y	Y	Y	Y	A	L	L	
Tiksi	5,063	S	CN	5.6	F	1			Y	Y	Y		Y	Y			C			
Igarka	6,183	S	RN	n.i.	G	0	Y	Y			Y	Y	Y	Y	Y	N	C			
Mezen	3,575	S	CN	3.9–4.5	G	3	Y	Y			Y	Y	Y	Y			C			
Belomorsk	11,217	S	CN	n.i.	F	4	Y	Y									C			
Rabocheostrovsk	2,900	S	CN	n.i.	F	5		Y	Y				Y				C		S	
Keret		S	CN	n.i.	F	0											C			
Bolshaya Piryu Guba		S	CN	n.i.	G	5						Y	Y	Y			C	S		
Ostrovnoy Gremikha	2,171	S	CN	n.i.	G	11				Y		Y	Y	Y			C	M		
Provideniya	1,970	S	CN	n.i.	G	2	Y	Y	Y	Y		Y	Y	Y	Y		C			
Ambarchik	4	S	CN	n.i.	F	0						Y	Y	Y			B			
Pevek	4,162	V	CN	13	G	0	Y	N	Y	Y		Y	Y				C			
Port Dikson	676	V	CN	15	G	1	Y	N	Y	Y		Y	Y	Y			C			

(continued)

Table 8 (continued)

Port	Population (2010)	Harbor size	Harbor type	Water depth (in m)[a]	Shelter	Tide range	ETA message	Pilot available	Fixed crane	Crane mobile	Crane float	Provisions	Water	Fuel/oil	Diesel	Deck supply	Repair code	Dry dock	Marine railway	Plans/comments
Ust Port		V	RN	n.i.	G	0											D			
Mys Novyy Port	1,790	V	RN	n.i.	F	2														
Nar Yan-Mar	21,658	V	RN	4.9	F	2	Y	Y		Y		Y	Y	Y			C			
Onega	21,359	V	RN	13.6	G	1	Y	Y				Y	Y				C			
Kovda	20	V	CN	n.i.	G	0	Y		Y											
Kandalaksha	35,654	V	RN	9.8	F	2	Y	Y		Y		Y	Y	N	Y		C		S	
Mys Abram		V	CN	n.i.	F	8			Y				Y							
Vitino		V	CN	11.1	P	2	Y	Y				Y	Y	Y	Y				S	
Varandey				14																Not listed in World Port Index
Amderma				2																Not listed in World Port Index
Sabetta				–																Port under construction
Kathanga				4.6							Y		Y	Y	Y					Not listed in World Port Index

Harbor Size: *L* large, *M* medium, *S* small, *V* very small, Columns containing Y or N: *Y* yes, *N* no

OR open roadstead, *CN* coastal natural, *RN* river natural, Shelter: *E* excellent, *G* good, *F* fair, *P* poor, *N* none

Water depth: *n.i.* not indicated

Repair Code: *A* major, *B* moderate, *C* limited, *D* emergency only, *N* none, Marine Railway: *L* large, *M* medium, *S* small

[a] Water depth (in m) is max. at berth

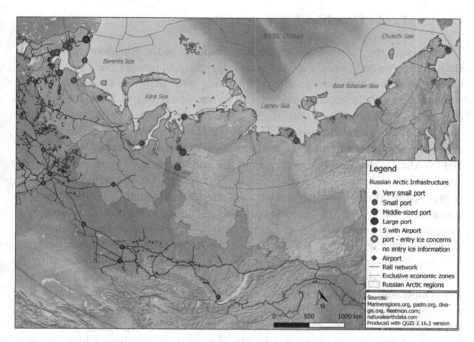

Fig. 9 Russian ports

port capacities that are limited in focus to natural resource extraction (Gunnarsson 2015, see also Staalesen 2012). A significant part of Arctic hydrocarbon resources is expected to be located in Russian territory (see Gunnarsson 2015) and NSR traffic from LNG cargo flows are expected to grow proportionally with the growth of Arctic hydrocarbon projects. As a result, transport of cargo could reach 100 million tons per year in 2030 (see Gunnarsson 2015) including transportation to European markets as well as Asian markets (if sanctions are removed). On the other hand, the fraction of international cargo transport between ports that lie outside Russia is still small (see Gunnarsson 2015). As stated in Gunnarsson (2015), Russian ports require modernization to be able to host and provide service to international traffic (see Ircha and Higginbotham 2016 and the references therein). Moreover, deep-water access as well as refuge and salvage support are needed. Currently, there are no Russian Arctic deep-water ports.

Regarding maritime infrastructure in the NSR, Gunnarsson (2015) points out the need for detailed structured analysis regarding the overall transport and logistics system of the NSR. This should include visualization of safety and security measures depicted on a map that includes physical infrastructure, communication, navigational systems, and response services. Data on commercial shipping in Arctic waters is crucial for insurance companies that need to determine operational conditions and risks for their assessments and in order to be willing to provide insurance coverage (see also Gunnarsson 2015).

Greenland and Faeroe Islands

Together with the Faeroe Islands, Greenland belongs to the Kingdom of Denmark. Both the Faroes and Greenland hold extensive rights and power of self-government, but Denmark controls foreign affairs, including most maritime concerns. A joint strategic plan for the Arctic was developed for the years 2011 to 2020 (see Ministry of Foreign Affairs Denmark et al. 2011). With the melting of the sea ice, the waters around Greenland and the Faeroe Islands are expected to experience an extensive increase in maritime traffic (see Ministry of Foreign Affairs Denmark et al. 2011). Already, for example, increasingly ice-free waters on Greenland's western coast are experiencing increased fishing for shrimp (Pandalus borealis) north of 66 degrees North (Pers. Comm., AnnDorte Burmeister, Greenland Institute for Natural Resources, March 2017). Tourism is also increasing (see Chap. 9 this volume). As shipping is a global industry, it is important to assure international high safety standards for navigation in the Arctic in order to prevent marine vessel accidents in a fragile environment such as the Arctic. On the one hand, ships navigating in the Arctic should be able to stand extreme weather and operational conditions, i.e., low temperatures and ice encounters as well as the risk of grounding in a vast area far away from ports (see Ministry of Foreign Affairs Denmark et al. 2011). This implies firstly that ships should bring their own rescue equipment and secondly, as other ships in the vicinity are highly likely first responders, which requires extensive information exchange and collaboration. Ports and their infrastructure play a vital role in enabling the capacity for this combination of independence and collaboration.

Greenland's maritime infrastructure is composed of many small (eight in the year 2016) and very small (16 in the year 2016) ports (see National Geospatial-Intelligence Agency 2016), but no medium or large ports; see Fig. 10 and Table 9 for an overview on ports and their infrastructure. These small ports link the country as roads between communities rarely exist. Greenland's shifting inland icecap and craggy coastal mountain terrain reduce the likelihood that roads will become a feasible transportation option.

The government of Greenland plans to explore its oil and mineral resources. These include a plan for five to ten large operative mines in the long run that will need port facilities for ore export (see Government of Greenland 2014). One mining company owning a license plans a deep-water port in Isua for ships up to 250 kt vessels. The project is 100% privately owned by the mining company that additionally plans this port in connection to their iron ore mining and processing plant also including a pipeline (see McCrae 2013). The project predicts year-round operations and the creation of 450 jobs, though the majority of the labor is likely to be imported.

One to two offshore oil and gas drilling projects are also expected to join the activity every second year from 2018. These will benefit from port structures with offshore support capabilities. Already, a large number of exclusive oil and gas licenses have been granted by Greenland's government. Though exploration in the 1970s did not recover oil, and no operating wells exist from that period (unlike

Table 9 Details on current and planned ports in Greenland and Faeror Islands (selection)

Port	Population (2010)	Harbor size	Harbor type	Water depth (in m)[a]	Shelter	Tide range	ETA message	Pilot available	Fixed crane	Crane mobile	Crane float	Provisions	Water	Fuel/oil	Diesel	Deck supply	Repair code	Dry dock	Marine railway	Plans/comments
Nuuk	17,316	S	CN	12.5	G	10	Y	Y		Y		Y	Y	Y	Y	Y	C		S	New container terminal with water depth of 15 m
Sisimiut	5,598	S	RN	7.8	F	4	Y	Y	Y	Y			Y		Y		B		S	
Aasiaat	3,102	S	CN	8.1	G	3	Y	Y		Y		Y	Y	Y			C		S	
Qeqertarsuaq	845	S	CN	n.i.	G	3	Y	N	N	Y		Y	Y	N	Y		C			
Uummannaq Harbor	~1300	S	CN	n.i.	F	0	Y	Y	Y	Y		Y	Y	N			C		S	
Kusanartoq		S	CN	n.i.	G	6	N	N					Y				D			
Paamiut (Frederikshab)	1,515	S	CN	n.i.	F	3	Y	Y	Y			Y	Y	Y	Y		C		S	
Illulissat (Jakobshavn)	4,541	S	CN	6.9	F	3	Y	Y		Y		Y	Y				C			
Kongshavn		V	CB	n.i.	G	0							Y	Y			C			
Vestmanna		V	CN	n.i.	E	5	N	Y				Y	Y	Y	Y		C			
Qaqurtoq	3,229	V	CN	n.i.	F	6	N	Y	Y	Y		Y	Y	Y	Y		C		S	
Narsarssuaq		V	CN	n.i.	F	3	Y	N	Y	Y		Y	Y	Y	Y		C			
Gronnedal (Kangilinnguit)		V	CN	n.i.	F	3	Y					Y	Y	Y	Y		C			
Manitsoq	2,670	V	CN	7.9	F	4	Y	Y		Y		Y	Y	Y	Y		C		S	

(continued)

Table 9 (continued)

Port	Population (2010)	Harbor size	Harbor type	Water depth (in m)^a	Shelter	Tide range	ETA message	Pilot available	Fixed crane	Crane mobile	Crane float	Provisions	Water	Fuel/oil	Diesel	Deck supply	Repair code	Dry dock	Marine railway	Plans/comments
Kangamiut	353	V	CN	17	G	12	Y						Y				C			
Kangerlussuaq	499	V	CN	3	F	10	Y	N		Y		Y	Y		Y		C			
Niaqornat	58	V	OR	n.i.	G	0	N	N					Y		N					
Kajalleq Upernavik		V	CN	n.i.	F	5		N		Y			N	N	N		C			
Upernavik	1,181	V	RN	6.2	F	2	Y	N	Y			Y	Y	Y	Y		C			
Qutdleq		V	CN	n.i.	G	0	N	N					Y	Y			N			
Kangerluarsoruseq		V	CN	2	G	4	Y	Y		Y		Y	Y	Y	Y		C			
Attu	226	V	RN	n.i.	F	0			Y	Y		Y	Y	N	Y		N			
Qasigiannguit-Christianshab	1,171	V	RN	8	G	3	N	Y	Y	Y		Y	Y	Y	Y		N			
Pituffik (Thule Air Base)		V	RN	n.i.	G	3	Y	Y		Y		Y	Y	Y	Y		C			

Harbor Size: *L* large, *M* medium, *S* small, *V* very small, Columns containing Y or N: *Y* yes, *N* no

Harbor Type: *OR* open roadstead, *CN* coastal natural, *RN* river natural, Shelter: *E* excellent, *G* good, *F* fair, *P* poor, *N* none

Water depth: *n.i.* not indicated

Repair Code: *A* major, *B* moderate, *C* limited, *D* emergency only, *N* none, Marine Railway: *L* large, *M* medium, *S* small

^{a}Water depth (in m) is max. at berth

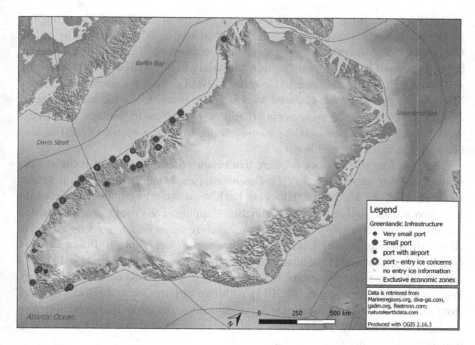

Fig. 10 Greenland ports

Alaska and Norway), a new exploratory phase has begun with an expanded range in both Eastern and Western Greenland (Geological Survey of Denmark and Greenland (GEUS) 2017).

While mining and oil and gas exploration are desirable for income - which among other things, might eventually allow full independence from Denmark - Greenland's home-rule authorities are aware that protecting their environment and society through sustainable development of their country is mandatory. A report prepared by the International Institute for Sustainable Development[3] analyzes the perspectives and expectations of stakeholders for sustainable development of Greenland taking into account environmental, social, and economic externalities. A special concern that needs to be addressed is how to deal with large investors that have great financial, industrial, and marketing powers, as in the mining sector. Similarly, there are significant complexities with respect to high expected levels of worker immigration and its effects on the rather small Greenlandic population of ca. 57,000 people (see International Institute for Sustainable Development 2013). The institute advises Greenland to consider lessons learned by countries with large migration and expatriate labor, such as Singapore or the United Arab Emirates. Greenland may also wish to heed lessons learned from the evolution of capital intensive infrastructural projects as mentioned in the section "Evolving Supply Provision".

---

[3] See also: http://www.iisd.org/.

Regarding maritime tourism in the Arctic waters of Greenland, the Kingdom of Denmark has improved port state control of cruise ships planning to sail to Greenland (see Ministry of Foreign Affairs Denmark et al. 2011). For instance, these ships need to report their position constantly to the "GREENPOS" reporting system under Greenland Command in order to avoid collisions and other accidents. Moreover, large ships are required to send their position via the satellite-based *long-range identification and tracking* (LRIT) system. New technology such as the satellite receipt of signals via the *automatic identification systems* (AIS) is also becoming available for monitoring.

The Kingdom of Denmark is aware that besides fisheries, tourism is the second most important export industry in Greenland (see Ministry of Foreign Affairs Denmark et al. 2011). This includes both land-based and cruise liner activities, with both also at least partly reliant on destination shipping. The government has created a transportation commission[4] in order to analyze transportation needs and give advice on how to develop Greenland's infrastructure according to future demand (see Bendsen et al. 2011; Ministry of Foreign Affairs Denmark et al. 2011). The current infrastructural situation of Greenland is mainly based on elderly military installations (see Bendsen et al. 2011) with two airports located rather far away from scattered populated locations. Significant distances make transportation very costly. A new container port in Nuuk is planned and initiated that will have a 320 m long berth with water depth of 15 m.

Challenges of Activities in the Arctic

Obstacles to reducing voyage time include the fact that there are large swathes of territory with low human habitation or opportunity for development. For example, along the Siberian coast between the Bering Strait and the port of Murmansk, long stretches of the landscape have geophysical constraints that do not permit stopovers or reliable communication and aid in case of emergencies (Verny and Grigentin 2009). This is confirmed by Ho (2010) who notes that additional investments in infrastructure and marine services including safety and security are needed before the NSR can be subject to continuous and large-scale shipping. An important consideration is the desire to minimize environmental impacts. This desire is actively voiced by all Arctic states, though levels of action may be disputed; see Arctic Resources and Transportation Information System (Arctis) Knowledge Hub (2017) for materials covering environmental impacts and the legal and regulatory frameworks for Arctic maritime activity affecting environmental quality. Ho (2010) as well as Smith and Stephenson (2013) state that several issues particularly need to be resolved for Arctic shipping, including:

---

[4]See also http://www.transportkommissionen.gl/.

- Access to reliable environmental observations for vessels operating in the Arctic regarding weather information including sea ice, wind, and ocean conditions and their forecasts,
- Provision of SAR services such as ice-breaker support services,
- Availability to hire, or direct provision of, Arctic-experienced work force to operate ships in the NSR,
- Enhancement in ship technology for independent ship operations in ice-covered waters,
- Installation of vessel traffic systems (VTS) on ships along narrow straits in the NSR to avoid accidents between ships,
- Integrated governance and regulatory framework based on the United Nations Convention on the Law of the Sea (see also Smith and Stephenson 2013),
- Ensure year-round supply logistics for companies working in the Arctic, e.g., drilling companies (see Higginbotham and Grosu 2014),
- Ensure fast response to accidents, e.g., oil spills or blowouts, with governmental support (see Higginbotham and Grosu 2014).

Interacting Demand and Supply: Outcomes

In the following sections, we analyze the economic attractiveness of the three new routes through the Arctic and their predicted future developments.

Economic Attractiveness of the Trans-Polar Route (TPR)

Though use of the TPR for Arctic shipping remains theoretical and is unlikely to become a reality in the near future, the potential displayed for viability of the route by mid-century indicates that long-term port investment decisions should consider the impact this route will have. If functioning, the TPR might serve effectively as a 'backstop' for trans-Arctic shipping that eliminates several economic and environmental challenges associated with either shore-based route concerns, such as taxes and fees from Russia or Canada, or running aground in poorly charted and/or narrow passages. Aksenov et al. (2016) suggest that by the middle of the century, the TPR may be as fast a route as the NSR, with either route taking approximately 11–16 days. Smith and Stephenson (2013) present similar findings regarding the viability of the TPR by mid-century, especially for Polar Class 6 vessels. These are polar vessels that are allowed to operate in summer/autumn ice conditions in medium first-year ice that might include old ice inclusions (see International Association of Classification Societies 2016). Since the TPR would not involve as many political risks and presents different safety risks from the shore-based route, competition from the route can be expected to be a favored choice under many future climate and economic outcomes, and to therefore limit the potential of the

NSR for trans-Arctic shipping. One significant uncertainty is how Arctic Ocean sea conditions, particularly wave actions, might change with reduced ice cover. Increased wave action would also affect poorly protected shorelines on the NWP and NSR routes through increased coastal erosion, however, and the net effect on the relative attractiveness of the routes is not currently predictable; see Emmerson and Lahn (2012) for more discussion of these risks.

The TPR is a substitute for shore-based Arctic shipping. Rather than a complement that furthers market integration and trade by connecting smaller ports to larger global trade, the opening of the TPR would effectively function as a highway bypass and again leave Arctic development to depend primarily on local forces.

Economic Attractiveness of the NSR

A feasibility study of regular container transport along the NSR is found in Verny and Grigentin (2009). This study analyzes the economic and technical feasibility of regular container transport along the NSR and concludes that, despite some challenges, at present the NSR constitutes a viable though peripheral alternative of transport between Asia and Europe. Since their research, additional studies, including Lee (2016) and Hansen et al. (2016), similarly stress both challenges and opportunities. They analyze whether the NSR has the potential to become a key axis of future shipping strategies or, whether the route is limited to being only a means of handling occasional overflow of the classical routes.

None of the studies find the NSR is currently viable as a key axis. Verny and Grigentin's optimistic analysis assumes that by the end of 2015, parts of the Arctic Ocean are navigable year-round, especially along the Russian coast. The long run viability of this route for cargo, however, hinges on additional concerns, in particular the competing role of the TPR. Verny and Grigentin (2009) state that the NSR reduces the voyages time by about 2,500 Nautical miles from North Asian ports such as ports of Japan, South Korea, and China, to Northwestern Europe north-range ports, e.g., Hamburg, Bremen, or Rotterdam. This translates to one-third of the maritime shipping time required for transport via the Suez Canal and approximately 10 days less voyage time. Verny and Grigentin (2009) conclude that, despite the higher costs for passing the NSR, such as fees for the authorization to travel the NSR imposed by the NSR Authority (NSRA), the NSR and the Trans-Siberian Railway, which carries more than 50% of Russia's foreign trade and transit freight (Russian Railways 2017), are roughly equivalent alternatives of transport between Asia and Europe.

Moreover, Ho (2010) states that the NSR above Russia between the North Atlantic and the North Pacific would trim the transit length about 5000 Nautical miles and a week's sailing time compared to alternative routes such as the Suez Channel combined with the Malacca Straits. However, cost savings are not linearly correlated with time and distance savings, so that the assessment of the NSR's attractiveness varies between great enthusiasm and complete disinterest. For

Table 10 Major NSR-located ports mainly Russia (Source: http://www.arcctic-lio.com/ (Accessed 19 Nov 2016))

Port	Location	Depth	Comments
Murmansk	Kola Bay on Barents Sea	10 m	Year-round ice free, ship repair
Kandalaksha	Kandalaksha Bay, White Sea	9.8 m	Year-round with ice-breaker support in winter
Vitino	Karelskiy shore of Kandalakshskaya Bay, White Sea	11.1 m	Year-round with ice-breaker support in winter
Onega	Near Onega Bay and Onega River, White Sea	13.6 m	May until January, ice-breaker upon request
Arkhangelsk	Near Dvina River, White Sea	9.2 m	Year-round with ice-breaker support in winter
Mezen	Mezen Bay, near Mezen River, White Sea	3.9–4.5 m	June until October
Naryan-Mar	Near Pechora River, Barents Sea	4.9 m	Mid June until Mid October or October with ice-breaker support
Varandey	Near Varandey Bay, Barents Sea	14 m	Year-round for ice-breaking-capable ships
Amderma	Kara Sea, east of Yugorskiy Strait	2.0 m	June until November
Sabetta	Western coast of Ob Bay	—m	Port under construction, part of Yamal LNG project
Dudinka	In the Yenisei River	11.8 m	Year-round with ice-breaker support
Dikson	Kara Sea near Yenisey River	15 m	Year-round with ice-breaker support
Khatanga	Laptev Sea at Lena River	4.6 m	Summer navigation only
Tiksi	Laptev Sea near Lena River	5.6 m	Mid July until Mid October
Pevek	In Chaunskaya Bay, Siberian Sea	13 m	Summer navigation only

instance, cost factors influencing the competitive advantage of the NSR include building costs for ice-classed ships, non-regularity of speeds including slower speeds, navigation challenges and increased risks, policy fees of insurers, and extra ice-breaker services and Russian NSR fees; see Liu and Kronbak (2010), Østreng et al. (2013), and Buixadé Farré et al. (2014).

Ircha and Higginbotham (2016) state that the NSR is more attractive for maritime shipping than the NWP as it is predicted to be ice-free year-round. The NWP allows for various routes to traverse the Arctic, but they include many small islands and narrow straits along the way, with shore-based ice formations and slower ice clearing during summer, so that navigation is more difficult than in the NSR. Regarding infrastructure, the NSR has well-distributed ports that can provide some service, shelter, and SAR capabilities (see Ircha and Higginbotham 2016 and Table 10).

The Russian part of the NSR is well-documented in terms of marine charts and navigational aids (see Ircha and Higginbotham 2016). Moreover, good communication and ice-breaking services are available on a year-round basis. SAR services are provided by Maritime Rescue Coordination Centers located at major Russian ports, i.e., Murmansk, Dikson, Tiksi, Prevek, and Provideniya.[5]

Russian ports are moving ahead of other Arctic regions in terms of safety and security, making both potential tourism and other economic and military activities less risky. The infrastructure of the Russian Arctic is notably higher than elsewhere. The larger population levels and longstanding political interest in Arctic development position the Russian Arctic and the NSR well for increased access and economic activity. In order for marine tourism and other nature-based maritime economic activity outside the state's purview, however, to increase in the Arctic, institutions, more than infrastructure, must shift in order to improve multi-use opportunities at ports (see Kaiser et al, Chap. 9 this volume). Instead, current private port development investments are focusing on single-use oil and gas resource extraction, such as the new port of Vitolo outside of Murmansk.

Economic Attractiveness of the NWP

The NWP faces major challenges to become a viable part of Arctic shipping activities (see Ircha and Higginbotham 2016). Despite the challenges of its navigability, there is an almost complete lack of infrastructure such as ports or places to serve as refuges for ships in distress (see Kives 2016). The Beaufort basin is identified as very important for the development of Alaska's and Canada's Arctic territories (see Higginbotham and Grosu 2014) due to oil, gas (especially LNG), and mineral resources, shipping, fishing, and tourism. Moreover, the port of the city of Adak located in Alaska could serve as an international hub for trans-Arctic shipping (see Ho 2010 and the references therein).

Despite the limitations of the route, the NWP has begun experimenting with trans-Arctic shipping. The *Nordic Orion* made a voyage in 2013 along the NWP. The voyage realized fuel and CO_2-emissions savings while simultaneously carrying 25% more cargo than is allowed by draft and weight for ships using the Panama Canal. This has triggered increased motivation and investment on both the eastern and western shores of the Arctic, though Russia and Scandinavia remain the most active countries (see Higginbotham and Grosu 2014).

Coordination and Governance

Currently, the NSR is administered by the Russian Marine Operations Headquarters (MOHQ) that has authorized two shipping companies to execute administration and control of the NSR (see Liu and Kronbak 2010). On the NWP route, the passage

---

[5]See also http://www.arctic-lio.com/nsr_searchandrescue.

is declared to be the Northwest Passage by the Canadian House of Commons in 2009. This sets up competition between the routes and complicates administration. In the Arctic, it is unlikely that cost reductions from competition can outweigh benefits from coordination that include risk mitigation to ships, local communities, and the environment, and economies of scope in services and safety (see Lasserre and Pelletier 2011). Internally in Russia, the increased and fragmented oversight also threatens the NSR-profitability for regular transit traffic though competition that results in prices below marginal cost (see Kaiser et al, Chap. 9 in this volume). Reorganizing this administrative set-up to achieve the gains suggested here is a significant challenge, however, because national control of the internal passages is key to sovereignty over other resources contained in Arctic waters (Lasserre and Pelletier 2011).

Examples of Bilateral and Multilateral Cooperation

There is great potential for cooperation with respect to maritime activities of the U.S. and Canada, although there are challenges (see Higginbotham and Grosu 2014). For instance, Nunavut is characterized by a very small Aboriginal population with self-governmental processes and fiscal dependence on the federal government that renders Arctic development and investment complex and lengthy (see Higginbotham and Grosu 2014). Any cooperation with U.S. counterparts faces additional levels of negotiation, extending beyond federal agreements. Furthermore, McCague (2014) states that the U.S. and Canadian governmental support for the Arctic is limited, while Russia is quite active, especially in building ice-breaking ships. Offshore oil negotiations are also important considerations in cooperation. While the Americans and Canadians have not bothered to settle their Arctic marine border, the Russians and Norwegians settled theirs finally in 2010. This was due to express desire to settle ownership of undersea oil reserves in order to begin exploration and exploitation of oil in the Barents Sea (see Kingdom of Norway 2010). This has led to cooperation on technical and environmental investments as well as joint ventures in exploration concerning both capital and labor under the Barents 2020 project (see DNV GL 2016).

The Inuit communities of the Canadian archipelago and Greenland are strongly tied by culture, language and outlook. Accordingly, Canada and Greenland (Denmark) have several relevant bilateral Memoranda of Understanding (MOUs) and other cooperative agreements on defense, cultural and educational arrangements, joint fisheries research, and marine environment (Government of Canada 2015). Moreover, Greenland, Iceland, and Norway share Danish and Nordic roots that facilitate communication and cooperation. Port activities are longstanding and the relatively open waters of Iceland and Norway have connected these nations to the broader North Atlantic for centuries. The countries are party to major international agreements that include regional fisheries management through the Northwest Atlantic Fisheries Organization (NAFO). NAFO and the Joint Norwegian-Russian

Fisheries Commission in the Barents are the only extant international Arctic fisheries agreements and will be important for fostering greater cooperation as climate change shifts species movement patterns.

Pan-Arctic cooperation generally occurs through the Arctic Council, which consists of the eight Arctic countries and representatives from the Arctic indigenous groups. The council has no direct regulatory power but works to secure multilateral cooperation and to support and use scientific efforts to improve a range of social, political, economic and environmental outcomes in the Arctic. With direct consequences for port development, the Council has succeeded in generating two general agreements (one regarding marine oil spill response and the other on search and rescue, each discussed below). These have been the two most successful collaborations of the Council at the policy level.

Safety and Security

Figure 11 shows the search and rescue agreement areas of the Arctic countries negotiated through the Arctic Council and pronounced in the Nuuk Declaration

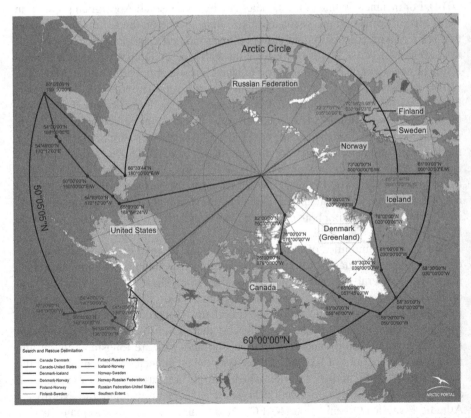

Fig. 11 SAR delimitations (Source: http://library.arcticportal.org/1474/11/search-rescue_ensku_110524_minnka.jpg)

(2011). The SAR agreement was the first multilateral agreement adopted by the Arctic Council. Adoption occurred at the Council's Foreign Ministers' Meeting in May 2011 in Nuuk (Ministry of Foreign Affairs Denmark et al. 2011). It set up expectations and commitments for collaborative safety and rescue activities within the Arctic.

The Polar Code supplements the SAR rules in the Arctic with additional rules on rescue equipment, fire fighting, and ice navigation in uninhabited areas. More precisely, it covers the full range of shipping related matters relevant to navigation in waters surrounding the two poles from ship design, construction, and equipment to operational and training concerns. It has been prepared by the IMO and came into force in January 2017.[6]

The code includes some mandatory matters for safety and pollution prevention as well as polar ship certification, classifying ships according to where they are allowed to travel due to ship design and equipment provision. Many of the provisions of the Polar Code are, however, voluntary. There is some concern that as this relatively weak regulation replaces no regulation, countries will cease to pursue higher standards of safety once they meet the minimum requirements. That is, having established low but clear standards, any efforts to increase them, and their costs, will be harder to initiate and agree upon as the regulated status quo can be cited as evidence of sufficient concern (see Lecraw 1984 for more on economic consequences of standards). This may become particularly worrisome for types of use that involve many untrained individuals in remote areas, such as with cruise ships.

Environmental Risks

The Agreement on Cooperation on Marine Oil Pollution Preparedness and Response in the Arctic, known also as the Kiruna Declaration (2013) was the second finalized formal policy outcome of Arctic Council work. Some of the agreement focuses on getting individual countries to engage in best practices in their own waters, but the agreement also makes provisions for sharing information about marine oil pollution that has international consequences, and for sharing direct response resources. Transboundary threats such as marine oil pollution require multilateral agreements. Legislation should be considered a vital component of marine use of the Arctic.

A focus of the recent U.S. Chairmanship of the Arctic Council (2015–2017) has been increased effort to prevent invasive species introductions and spread in the Arctic. Together with Arctic Council working groups CAFF (Conservation of Arctic Flora and Fauna) and PAME (Protection of the Arctic Marine Environment), the U.S. and Norway are developing a more coordinated approach to invasive species prevention. This is due for public presentation at the Arctic Council Ministerial in Spring 2017 (Conservation of Arctic Flora and Fauna (CAFF) working group 2016). Increased invasive species monitoring and awareness is particularly important as the

---

[6]See also http://www.imo.org/en/mediacentre/hottopics/polar/pages/default.aspx, last call 11 Dec 2016.

warming climate makes establishment of new species more likely and as increased traffic in the Arctic increases the propagule pressure of introductions.

The challenges here for detection and prevention are significant (Kourantidou et al. 2015). There is thus an important role for port infrastructure and management. Ports serve as concentrated points of entry and all aspects of construction and design should work to minimize successful establishment and/or maximize early detection and rapid response options should species successfully establish. This requires integrated consideration of the maritime routes and the potential biological and ecological threats along them. See Holbech et al, Chap. 7 this volume for additional information on governance activities for invasive species in Arctic waters.

The Role of Military Use in Port Development

As climate changes and the opening of Arctic waters increase Arctic maritime activities, security issues also rise, e.g., cases of authorized or unauthorized entry into U.S. American, Canadian, Danish (Greenlandic and Faeroese), Icelandic, and/or Russian waters (see Ircha and Higginbotham 2016). There are several programs to enhance off-shore patrolling and ice-breaking ship presence from the Canadian government. Moreover, NATO has declared the Arctic as a strategically important region (see Gabriel 2010), where northern member nations perform individual or collaborative operations. The U.S. Navy is increasing training efforts for their personnel regarding northern territories and operations including submarines, surface ships, and aircraft (see Ircha and Higginbotham 2016). NATO and Russia are increasingly wary of military or other escalations in the area (Chamberlain 2013). This activity at times involves considerable port activity. In particular, Russia is increasing its permanent military presence in the Arctic (Tomkiw 2016).

Military activities may increase tensions in the area, and they increase uncertainty for other activities in the region. This potentially will reduce economic and social activity. At the 40th annual meeting of the Joint Fisheries Commission between Norway and Russia (2016), for example, for the first time members of the Russian Northern Fleet were included. Further, Norwegian research expeditions that have occurred regularly for 40 years will now require a representative of Russia's Department of Defense on-board (Staalesen 2016).

Military activities have historically been subject to lower levels of environmental standards and higher degradation from increased activities may be expected (Josephson 2016). In Greenland, for example, toxic waste from WWII military activity that was expected to remain frozen into the far indefinite future may be thawing and causing damages by the end of the century (Martirosyan 2016). Decades of Russian nuclear testing in the Arctic will have lingering impacts for generations to come (Guruswamy and Aamodt 1999)

Historically, military investments in the Arctic have had other types of long term impacts as well, particularly from infrastructure investments. For instance, WWII left Northern Norway in a shambles, but also added roads and port infrastructure meant for rapid military movement and extraction of Norway's natural resources

(Hunt 2015) that remain integral components of northern infrastructure today. The needs and investments of the military today will shape much regarding infrastructure of the future.

Conclusions

This chapter has taken a local and regional view of port development in the Arctic. The idea that outcomes for Arctic development should depend more on local and regional choices than global forces is potentially contentious. We argue here, however, that with respect to increased utilization of the three potential trans-Arctic marine routes (the NSR, the NWP, and the TPR), local and regional choices for local development will greatly affect the infrastructure investments in different portions of the Arctic. This is in part because many of the global forces at work (e.g. trans-Arctic shipping, trade and resource extraction) are set to blow right past Arctic opportunities. The TPR is likely to prove superior to either shore-based route, and resource extraction is too volatile an industrial sector upon which to build resilient healthy communities. The time to a viable TPR route is short and estimates of the time left until its viability continue to shrink as we gain more understanding of the continuing sea-ice loss in the Central Arctic Ocean.

The shore based routes of the NWP and the NSR differ significantly in their current and expected uses. Indigenous uses, including subsistence fishing and hunting, and a slowly developing cruise tourism industry are key factors in the NWP. These depend heavily on the state of the ecosystems and local decisions of American, Canadian and Greenlandic indigenous users. Resource extraction interests in the three countries, and shifts in accessibility as land routes are rendered impassable by melting permafrost, may increase pressure for ports that facilitate these extractive industries. The NSR, with a much higher population overall, and a much lower percentage of indigenous populations, tends more heavily to resource extraction, cargo shipping and military use driven by Russian national interests. It is more economically competitive in this arena than the NWP.

There is increasing emphasis on sustainable development in these regions. This stems from the realization of high social costs of extractive industry in the north not only due to the well-known cyclical challenges of resource industries, but also due to societal impacts including gender and income disparities (see Kaiser et al, Chap. 9 this volume) and increased local climate change impacts through, e.g., black carbon and the introduction of damaging invasive species (see Holbech et al, Chap. 7 this volume).

Port development along the two routes should be expected to follow these divergent paths, unless governance and institutions at the local, national, and international levels evolve quickly to shift these interests. Signs of such shifts in uses of the NSR are nascent but not secure. Development efforts for nature tourism are seen as valuable options but there is little knowledge infrastructure or willingness to make compromises needed to facilitate international visitors (see Kaiser et al, Chap. 9, this volume). Nor is there willingness to either risk profits from oil and gas extraction or limit flexibility of military operations.

On both shore-based routes, concerns about the environment have been taken more seriously in some dimensions than in others. Since ports are a point of entry and focus of many human-dispersed environmental costs, their design and infrastructure must take into consideration these challenges. This requires cooperation with neighbors as well as more distant trading partners.

Multilateral cooperation in the Arctic has been fostered for 20 years by the Arctic Council. Successful international agreements negotiated through the Council, while few in number, have critical influence on Arctic port infrastructure decisions for the future. The Nuuk 2011 and Kiruna 2013 Declarations, on Search and Rescue and Marine Oil Pollution respectively, directly impact port infrastructure requirements. The IMO's Polar Code also introduces standards and regulation at the international level.

Local demands and logistical challenges to supply should be primary drivers of public port infrastructure development in the Arctic. Investment for resource extraction has been forthcoming from private enterprise; public oversight of this investment can argue for broader public interests and improve social outcomes. The public's broader community goals require direct, inclusive, public planning and investment to avoid inefficient redundancies, as spatial, seasonal, and human and physical capital limitations crowd viable economic activities into a few locations.

References

Aarluk Consulting, Gartner Lee Limited, & Anderson, C. (2005). Strategic plan for the Iqaluit deepwater port project. Internet Source: http://www.tunngavik.com/documents/publications/2005-08-00%20Iqaluit%20Deepwater%20Port%20Strategic%20Plan.pdf. Last call: 07 Dec 2016.

Aksenov, Y., Popova, E. E., Yool, A., Nurser, A. J. G., Williams, T. D., Bertino, L., & Bergh J. (2016). On the future navigability of Arctic sea routes: High-resolution projections of the Arctic Ocean and sea ice. Marine Policy, 75, 330–317.

Arctic Council. (2009). Arctic marine shipping assessment 2009 report. Internet Source: http://www.pmel.noaa.gov/arctic-zone/detect/documents/AMSA_2009_Report_2nd_print.pdf. Last call: 14 Nov 2016.

Arctic Resources and Transportation Information System (Arctis) Knowledge Hub. (2017). Marine transport and logistics. Internet Source: http://www.arctis-search.com/Marine+Transport+and+Logistics. Last call: 26 Mar 2017.

Barton, K. (2016). Crews connect Inuvik to Tuktoyaktuk highway in the middle. Internet Source: http://www.cbc.ca/news/canada/north/inuvik-tuktoyaktuk-highway-1.3526669. Last call: 19 Nov 2016.

Bendsen, S., Nordskilde, J., & Jensen, M. P. (2011). The transport commission of Greenland. Internet Source: http://abstracts.aetransport.org/paper/index/id/3692/confid/17. Last call: 11 Dec 2016. ISSN:2313-1853.

Bremenports. (2013). Iceland intends to build a new port on the Arctic Ocean and wishes to cooperate with the planning experts at Bremenports. Internet Source: http://www.bremenports.de/1267_2&template=print. Last call: 13 Nov 2016.

Buixadé Farré, A., Stephenson, S. R., Chen, L., Czub, M., Dai, Y., Demchev, D., Efimov, Y., Graczyk, P., Grythe, H., Keil, K., Kivekas, N., Kumar, N., Liu, N., Matelenok, I., Myksvoll, M., O'Leary, D., Olsen, J., Pavithran, S. A. P., Petersen, E., Raspotnik, A., Ryzhov, I., Solski, J., Suo, L., Troein, C., Valeeva, V., van Rijckevorsel, J., & Wighting, J. (2014). Commercial Arctic shipping through the Northeast Passage: Routes, resources, governance, technology, and infrastructure. *Polar Geography, 37*(4), 298–324. http://www.enr.gov.nt.ca/state-environment/73-trends-shipping-northwest-passage-and-beaufort-sea.

Chamberlain, N. (2013). Increasing military activity in the Arctic. Internet Source: http://natowatch.org/sites/default/files/briefing_paper_no.32_-_increase_in_military_activity_in_the_arctic.pdf. NATO Watch Briefing Paper No. 32; Last call: 19 Jan 2017.

Chambers, S. (2015). Iceland's president calls for coordinated Arctic port development. Internet Source: http://splash247.com/icelands-president-calls-for-coordinated-arctic-port-development/. Last call: 12 Nov 2016.

Conservation of Arctic Flora and Fauna (CAFF) working group. (2016). Arctic invasive species. Internet Source: http://www.caff.is/invasive-species. Last call: 19 Jan 2017.

Dasgupta, S. (2016). What are deep water ports? Internet Source: http://www.marineinsight.com/ports/what-are-deep-water-ports/. Last call: 19 Jan 2017.

DeMarban, A. (2015). Work toward deep-water port in Alaska Arctic on hold, Army Corps says. Internet Source: https://www.adn.com/arctic/article/corps-deepwater-port-us-arctic-hold/2015/10/26/. Last call: 19 Jan 2017.

Dlouhy, J. A. (2017). Trump weighing Eni's Arctic drilling bid in post-Obama pivot. Internet Source: https://www.bloomberg.com/news/articles/2017-03-16/trump-weighing-eni-bid-to-drill-in-arctic-waters-after-obama-ban. Last call: 23 Mar 2017.

DNV GL. (2016). Barents 2020 reports. Internet Source: https://www.dnvgl.com/oilgas/arctic/barents-2020-reports.html. Last call: 16 Dec 2016.

Ducharme, S. (2016). Iqaluit deep sea port on schedule, Nunavut official says – City councillors want more consultation on port, small craft harbour. Internet Source: http://www.nunatsiaqonline.ca/stories/article/65674iqaluit_deep_sea_port_on_schedule_nunavut_official_says/. Last call: 17 Jan 2017.

Emmerson, C., & Lahn, G. (2012). Arctic opening: Opportunity and risk in the high north, Lloyd's report. Internet Source: https://www.lloyds.com/~/media/files/news%20and%20insight/360%20risk%20insight/arctic_risk_report_webview.pdf. Last call: 26 Mar 2017.

Fletcher, S., Robertson, T., Higman, B., Janes, M., Sughroue, A., DeCola, P., & Mom, C. (2016). Bering sea vessel traffic risk analysis. Internet source, Dec 2016 ocean conservancy report by Nuka Research and Planning Group. http://www.oceanconservancy.org/places/arctic/bering-sea-vessel-traffic.pdf. Last call: 25 Mar 2017.

Frizzel, S. (2017). Public gets first look at plans for Iqaluit's deep sea port. CBC News. http://www.cbc.ca/news/canada/north/iqaluit-deepwater-port-design-plans-1.4008680. Last call: 25 Mar 2017.

Gabriel, D. (2010). Nato Arctic security and Canadian sovereignty in the far north. Global research. Internet Source: http://www.globalresearch.ca/nato-arctic-security-and-canadian-sovereignty-in-the-far-north/22048. Last call: 20 Nov 2016.

Geological Survey of Denmark and Greenland (GEUS). (2017). Oil and gas Greenland. Internet Source: http://www.geus.dk/UK/energy/oil-gas-gl/Pages/default.aspx. Last call: 25 Mar 2017.

Government of Canada. (2015). International Arctic partners. Internet Source: http://www.international.gc.ca/arctic-arctique/partners-international-partenaires.aspx?lang=eng. Last call: 19 Jan 2017.

Government of Greenland. (2014). Greenland's oil and mineral strategy 2014–2018. Internet Source: http://naalakkersuisut.gl/~/media/Nanoq/Files/Publications/Raastof/ENG/Greenland%20oil%20and%20mineral%20strategy%202014-2018_ENG.pdf. Last call: 09 Dec 2016.

Gramer, R. (2017). Trump's military buildup threatens to gut U.S. coast guard. Foreign policy. Internet Source: http://foreignpolicy.com/2017/03/02/trump-military-build-up-threatens-to-cut-u-s-coast-guard-budget-cuts-department-of-homeland-security/. Last call: 23 Mar 2017.

Gunnarsson, B. (2015). Future development for the northern sea route. Internet Source: http://www. maritime-executive.com/editorials/future-development-of-the-northern-sea-route. Last call: 19 Jan 2017.

Guruswamy, L. D., & Aamodt, J. B. (1999). Nuclear arms control: The environmental dimension. *Colorado Journal of International Environmental Law and Policy, 10*, 267–318.

Hansen, C. Ø., GrØnsedt, P., Graversen, C. L., & Hendriksen, C. (2016). *Arctic shipping.* Copenhagen: CBS.

Higginbotham, J., & Grosu, M. (2014). The Northwest Territories and Arctic maritime development in the beaufort area. Centre for International Governance Innovation – Policy Briefs. https://www.cigionline.org/publications/northwest-territories-and-arctic-maritime-development-beaufort-area. Last call: 19 Jan 2017.

Ho, J. (2010). The implications of Arctic sea ice decline on shipping. *Marine Policy, 34*(3), 713–715.

Hunt, V. (2015). *Fire and ice: The Nazi's scorched earth campaign in Norway.* Copenhagen: The History Press.

International Association of Classification Societies. (2016). Requirements concerning polar class. Internet Source: http://www.iacs.org.uk/document/public/Publications/Unified_requirements/PDF/UR_I_pdf410.pdf. Last call: 23 Mar 2017.

International Institute for Sustainable Development. (2013). Sustainable development in Greenland: Perspectives from a preliminary stakeholder consultation. Internet Source: https://www.iisd.org/pdf/2014/greenland_sustainable_development.pdf. Last call: 07 Dec 2016.

Ircha, M. C., & Higginbotham, J. (2016). Canada's Arctic shipping challenge, Chapter 15. In A. K. Y. Ng, A. Becker, S. Cahoon, S. -L. Chen, P. Earl, & Z. Yang (Eds.), *Climate change and adaptation planning for ports* (Routledge studies in transport analysis, pp. 232–245). New York: Routledge.

Jenkins, A. (2017). Spanish oil repsol company makes biggest U.S. onshore find in 30 years. Internet Source: http://fortune.com/2017/03/10/oil-discovery-repsol/. Last call: 23 Mar 2017.

Jepson, T. (2017). The world's most dangerous cruise? 1,070-capacity ship takes on the Northwest Passage. The Telegraph. http://www.telegraph.co.uk/travel/cruises/news/worlds-most-dangerous-cruise-arctic-northwest-passage/. Last call: 25 Mar 2017.

Josephson, P. R. (2016). Russia, state capitalism and Arctic degradation. *Global Environment, 9*(2), 376–413.

Kingdom of Norway. (2010). Treaty between the kingdom of Norway and the Russian Federation concerning maritime delimitation and cooperation in the Barents sea and the Arctic Ocean. Internet Source: https://www.regjeringen.no/globalassets/upload/SMK/Vedlegg/2010/avtale_engelsk.pdf. Last call: 16 Dec 2016.

Kiruna (2013). Kiruna declaration. The eight ministerial meeting of the Arctic council, Kiruna, 15 May 2013. Internet Source: https://oaarchive.arctic-council.org/handle/11374/93. Last call: 16 Oct 2017.

Kives, B. (2016). What the closure of an Arctic seaport in Manitoba could mean for Canadian sovereignty – closure of Canada's only deepwater mainland Arctic port may pose a problem. Internet Source: http://www.cbc.ca/news/canada/manitoba/churchill-port-analysis-1.3696199. Last call: 19 Nov 2016.

Kourantidou, M., Kaiser, B. A., & Fernandez, L. (2015). Towards Arctic resource governance of marine invasive species. Internet Source: http://www.arcticyearbook.com/images/Articles_2015/9.Towards-Arctic-Resource.pdf. Arctic yearbook 2015; Last call: 19 Jan 2017.

Lasserre, F., & Pelletier, S. (2011). Polar super seaways? Maritime transport in the Arctic: An analysis of shipowners' intentions. *Journal of Transport Geography, 19*(6), 1465–1473.

Lecraw, J. D. (1984). Some economic effects of standards. *Applied Economics, 16*(4), 507–522.

Lee, S. W. (2016). Necessary conditions for the commercialization of Arctic shipping. In *Challenges of the changing Arctic: Continental shelf, navigation and fisheries.* Copenhagen: Brill Nijhoff. http://www.brill.com/products/book/challenges-changing-arctic#TOC_1. Last call: 23 Mar 2017.

Liu, M., & Kronbak, J. (2010). The potential economic viability of using the Northern Sea Route (NSR) as an alternative route between Asia and Europe. *Journal of Transport Geography, 18,* 434–444.

Martirosyan, L. (2016). Waste from a secret US base was left to be entombed "forever" in Greenland's ice. but forever has changed. Internet Source: http://www.pri.org/stories/2016-08-11/entombed-cold-war-era-us-military-base-may-be-exposed-climate-change. Last call: 19 Jan 2017.

McCague, F. (2014). Arctic shipping route develop at top of the world. *Cargo Business News,* 92(4), 18–22.

McCrae, M. A. (2013). Greenland iron ore mine gets green light. Internet Source: http://www.mining.com/greenland-iron-ore-mine-gets-gets-a-green-light-25569/. Last call: 07 Dec 2016.

McDowell Group. (2016). Port of Nome – strategic development plan. Internet Source: http://www.mcdowellgroup.net/wp-content/uploads/2016/04/1555-Port-of-Nome-Strategic-Development-Plan-Final.pdf. Last call: 14 Nov 2016.

Ministry of Foreign Affairs Denmark, Department of Foreign Affairs Greenland, & Ministry of Foreign Affairs Faroes. (2011). Kingdom of Denmark strategy for the Arctic 2011–2020. Internet Source: http://www.uniset.ca/microstates/mss-denmark_en.pdf. Last call: 10 Dec 2016. ISBN:561-5.

National Energy Authority of Iceland (Orkusofnun). (2017). Oil and gas exploration. Internet Source: www.nea.is/oil-and-gas-exploration/. Last call: 25 Mar 2017.

National Geospatial-Intelligence Agency. (2016). *World port index* (25th ed.). Internet Source: http://msi.nga.mil/MSISiteContent/StaticFiles/NAV_PUBS/WPI/Pub150bk.pdf. Last call: 19 Jan 2017.

Nuuk (2011). Nuuk declaration. The seventh ministerial meeting of the Arctic council, Nuuk, 12 May 2011. Internet Source: https://oaarchive.arctic-council.org/handle/11374/92. Last call: 16 Oct 2017.

Østhagen, A. (2015). Coastguards in peril: A study of Arctic defence collaboration. *Defence Studies, 15*(2):143–160.

Østreng, W., Eger, K. M., Fløistad, B., Jørgensen-Dahl, A., Lothe, L., Mejlænder-Larsen, M., & Wergeland, T. (2013). *Shipping in Arctic Waters: A comparison of the Northeast, Northwest and Trans Polar Passages.* Berlin/Heidelberg: Springer.

Pearce, T. D., Smit, B., Duerden, F., Katayoak, F., Inuktalik, R., Goose, A., & Wandel, J. (2008). Travel routes, harvesting and climate change in Ulukhaktok, Canada. In *The borderless North. Northern Research Forum.* Oulu: The Thule Institute, University of Oulu. https://www.rha.is/static/files/NRF/OpenAssemblies/Oulu2006/nrf_publication_4th_procedings.pdf.

Ragner, C. L. (Ed.). (2000). The 21st century – Turning point for the Northern Sea Route? In *Proceedings of the Northern Sea Route User Conference*, Oslo, 18–20 Nov 1999. Springer.

Russian Railways. (2017). Russian railways website. Internet Source: http://eng.rzd.ru/statice/public/en?STRUCTURE_ID=4314. Last call: 26 Mar 2017.

Smith, L. C., & Stephenson, S. R. (2013). New trans-Arctic shipping routes navigable by midcentury. *PNAS, 110*(13), 1191–1195.

Staalesen, A. (2012). In Russian Arctic, a new major sea port. Internet Source: http://barentsobserver.com/en/energy/russian-arctic-new-major-sea-port-06-08. Last call: 11 Dec 2016.

Staalesen, A. (2016). Security interests spice up Barents fishery talks. Internet Source: http://thebarentsobserver.com/en/industry-and-energy/2016/10/security-interests-spice-barents-fishery-talks. Last call: 19 Jan 2017.

Stephenson, S. R., Smith, L. C., & Agnew, J. A. (2011). Divergent long-term trajectories of human access to the Arctic. *Nature Climate Change, 1,* 156–160.

Tomkiw, L. (2016). Russia military increases Arctic permanent presence amid regional tension. Internet Source: http://www.ibtimes.com/russia-military-increases-arctic-permanent-presence-amid-regional-tension-2285782. Last call: 19 Jan 2017.

United States White House. (2017a). President Trump is working to rebuild our nation's infrastructure. Internet Source: https://www.whitehouse.gov/the-press-office/2017/02/28/president-trump-working-rebuild-our-nations-infrastructure. Last call: 23 Mar 2017.

United States White House. (2017b). Budget. Internet Source: ttps://www.whitehouse.gov/omb/budget. Last call: 23 Mar 2017.

United States White House. (2017c). An America first energy plan. Internet Source: https://www.whitehouse.gov/america-first-energy. Last call: 23 Mar 2017.

U.S. Army Corps of Engineers. (2013). Alaska deep-draft Arctic port system study. Internet Source: http://www.poa.usace.army.mil/Library/Reports-and-Studies/Alaska-Regional-Ports-Study/. Last call: 14 Nov 2016.

Van Dusen, J. (2016). Iqaluit's deep sea port inches forward. Internet Source: http://www.cbc.ca/news/canada/north/iqaluit-deep-sea-port-1.3782182. Last call: 07 Dec 2016.

Verny, J., & Grigentin, C. (2009). Container shipping on the northern sea route. *International Journal of Production Economics, 122*, 107–117.

Wald, E. R. (2016). Alaska's 10 billion barrel oil discovery: What you need to know. Internet Source: https://www.forbes.com/sites/ellenrwald/2016/10/06/alaskas-10-billion-barrel-oil-discovery-what-you-need-to-know/#11efba2b47ef. Last call: 23 Mar 2017.

Arctic Ports: Local Community Development Issues

Brooks A. Kaiser, Julia Pahl, and Chris Horbel

Abstract Climate and economic forces are both transforming Arctic communities. Restructured governance of marine transportation and community development investment should work to promote economic growth and development within frameworks that accommodate sustainable resource use and community cultures. Marine ports are vital community links and components of this infrastructure. This chapter discusses historical lessons as well as Arctic community demands initiating from resource extraction, tourism, fishing, and culture for successful port development.

Decisions over both port infrastructure and industry regulation are crucial to ensure sustainable economic development and social equality while avoiding overcapitalization. We find that Arctic fisheries are currently well regulated, whereas resource extraction has the most risk of creating negative social costs for communities. Arctic marine tourism is profitable and growing, but challenges include high demand elasticity for Arctic cruises as well as imbalances in what the local population can provide to meet the highly seasonal demand. Cooperation within and across Arctic nations will be required to meet these challenges and realize important goals of Arctic port development, including social equality, conservation of living marine resources and reduction of overcapitalization.

Keywords Arctic economic development • Arctic community development • Arctic marine resource extraction • Arctic marine tourism • transportation networks

B.A. Kaiser (✉) • C. Horbel
Department of Sociology, Environmental and Business Economics, University of Southern Denmark, Niels Bohrs Vej 9, DK-6700 Esbjerg, Denmark
e-mail: baka@sam.sdu.dk; horbel@sam.sdu.dk

J. Pahl
SDU Engineering Operations Management, Department of Technology and Innovation, University of Southern Denmark, Campusvej 55, DK-5230 Odense, Denmark
e-mail: julp@iti.sdu.dk

© Springer International Publishing AG 2018
N. Vestergaard et al. (eds.), *Arctic Marine Resource Governance and Development*,
Springer Polar Sciences, https://doi.org/10.1007/978-3-319-67365-3_9

Introduction

Climate change is reducing ice cover on land and sea, and melting permafrost in the Arctic. This is increasing the pressures on economic opportunities for Arctic communities that are already shifting due to globalization. The resulting transformation of transportation options and needs will be significant. Furthermore, the transformations will be endogenous to decisions regarding access to Arctic communities. Port infrastructure and other investment decisions will be made that have lasting impacts on where and how economic development in the Arctic will proceed (Newton et al. 2016).

This chapter is a companion piece to Chap. 8 on port infrastructure. Arctic marine communities have varied uses for ports. These uses encompass resource extraction, tourism, fishing and cultural heritage. Dependence on destination shipping (i.e., imported cargo to and exported natural resources from northern communities) is also important in most locations and has grown with globalization. Decisions on port infrastructure investment need to respond to climate change and they must be made under highly uncertain economic futures. This chapter uses past experience with large scale infrastructure development in North America to elicit useful lessons for the region's continuing development.

Port infrastructure decisions should consider the complex economic and environmental uncertainties ahead in the Arctic. Historical context, gathered from other large scale and cumulative transport infrastructure decisions, provides some guidance over these uncertainties. A useful analytical question is, how the boom and bust of railroad expansion in the U.S. can be used as an analogy in terms of the lessons to be drawn for port infrastructure investments. Particularly insightful lessons can be drawn by considering how different rates of change in accessibility (as conditions vary for changing sea- and land-fast ice across ports) will affect the optimal investment patterns across the two shore-based routes. These differential changes will interact with the different end uses (oil and gas, mining, fishing, tourism, shipping) to affect Arctic communities. The two shore-based maritime routes traversing the Arctic marine environments are the Northern Sea Route (NSR), which crosses from the North Atlantic and Barents Sea into the North Pacific along Russian shores, and the Northwest Passage (NWP), which crosses from the North Pacific to the North Atlantic along American, Canadian and Danish (Greenlandic) shores. The (as yet hypothetical) Trans Polar Route (TPR) crosses the Central Arctic Ocean in the middle.

The TPR presents a more complicated aspect of the story, and parallels more closely the impact of highway bypasses on cities than it does the completion of the transcontinental railroad. This chapter does not address the TPR further than to note that it serves as a limit to ambitions for Arctic activities based on trans-Arctic shipping. The three routes present very different economic, social, political, safety, and environmental concerns that will shift differently with climate changes.

As such, they present very different opportunities as well. Furthermore, the impacts of the differential rates of change may also vary with differences in institutional structures, policy, and governance decisions.

Arctic ports are typically anticipated to provide economic opportunity by servicing resource extractors. These extractive industries include both non-renewables such as mining and oil and gas, and renewables such as fishing (see Jørgensen-Dahl and Wergeland 2013). However, short seasons and limited and volatile demand could result in the same sort of over-capitalization seen in open access resource extraction directly. In such cases, stakeholders, e.g., fishermen with time-limited (derby style) harvest windows, may over-invest in capital (e.g., vessels) and infrastructure that might become idle or stranded much of the time. Changes to the economic landscape of local communities in such a setting could have far-reaching impacts. A discussion of the impacts of historical extraction of whale and walrus (renewable resources) provides insight into these challenges.

The chapter focuses on the shore based routes comparatively and the different conditions that these face for the development of the communities. The different opportunities that they provide (especially for destination shipping, resource extraction, Arctic tourism and fishery) and the economic, social, political, safety and environmental consequences that arise will be discussed. This will also include cross-industry considerations. The sparsely populated NWP remains a too-costly alternative for trans-Arctic shipping, while the NSR is becoming increasingly viable due to less ice. This advantage to the NSR is only expected to last as long as the TPR is not also viable. Resource extraction of both renewable and non-renewable resources provides a strong driver of activity on both the NSR and NWP routes. The NWP has higher percentages of indigenous populations and greater emphasis on tourism, so that the number of social conflicts being addressed directly is currently higher along the NWP. The outcomes for these conflicts will be reflected in port decisions that can be expected to create lasting impacts on the direction of change in Arctic marine community development.

Effects of Climate Change on Arctic Maritime Activity

The Arctic is especially sensitive to fluctuating and warming temperatures; see, e.g., the report of the Intergovernmental Panel on Climate Change (2001). In fact, the Arctic is experiencing warming at a greater rate than any other parts of the planet. This has been the case for many years (see Johannessen et al. 1999). Moreover, the rate of change in the Arctic is expected to continue to diverge from the global rate of change. This will intensify the effects and hasten warming further (see, e.g., Koenigk et al. 2013). Scientists continue to find new indications of this change that surpass their forecasts. In 2016, the amounts of difficult-to-navigate multi-year ice fell dramatically in volume and coverage compared to expectations (see Richter-Menge et al. 2016). This changing landscape in the Arctic increasingly creates questions and concerns about navigational safety. Most recently, Aksenov et al.

(2016) have integrated climate forecasts for ice, wind, and other climate variables to estimate how climate changes will physically change the possibilities of Arctic routes for navigation.

The Arctic Communities

Arctic communities often have or desire connections to global economic activity that are fraught with challenges. These connections and the communities themselves are often little understood by outsiders (Nuttall 2012). There exist multiple definitions of the Arctic, so that population estimates vary from about 400,000 to 4,000,000 people.

Figure 2 illustrates how this population is distributed. The sparse dot densities indicate low population densities overall, with extremely sparse populations from Eastern Russia to Greenland along the NWP. Large portions of the population are small indigenous subsistence communities which are not well incorporated into global market economies. Along the NWP, American, Canadian and Greenlandic (Danish) Inuit communities have long-standing trade and communications, particularly in the winter months, over ice; see Kaiser and Parchomenko, Chap. 6 this volume. In Iceland, Norway and western Russia, relatively ice-free waters and the effects of the Gulf Stream have produced long-standing coastal communities that rely heavily on local and commercial fisheries. Distances remain great and marine transportation is vital. In central Arctic Russia, for centuries efforts to develop the region have involved transplanting people from southern communities in order to build infrastructure and develop resource extraction. The region is currently heavily investing in oil and gas exploration and development. To enable this, they are importing transient labor at high rates, with attendant community conflicts (Saxinger 2016).

Figure 1 illustrates some of the demographics of the Arctic. Regional data compiled under the ECONOR project (Glomsröd and Aslaksen 2008) are mapped and show the percentage of the population that is aboriginal as well as area population densities. Port information is also included to show the size and ice conditions of existing ports. Much of the total population lives in Norway and western Russia, where climate has been more moderate than other parts of the Arctic. The Barents Sea waters are now virtually ice free most of the year. Murmansk is the largest regional port. It connects Russia to its close neighbor to the west, Norway, and on the eastern side, to the NSR.

The low levels of population and infrastructure along the NWP, as well as stretches of the NSR, are also obvious. One can see that the NSR is likely to exhibit quite different demands from their ports than the NWP. In particular, differences in indigenous population shares should be expected to result in significantly different outcomes for social welfare from any increases in port development and use that are based primarily on resource extraction.

The Arctic is not, as many imagine, an empty frontier (Steinberg 2015); its populations have long adapted to the climate and seasonal constraints successfully, with considerable circumpolar trade (Aporta 2009; Morrison 1991; Stuhl 2016).

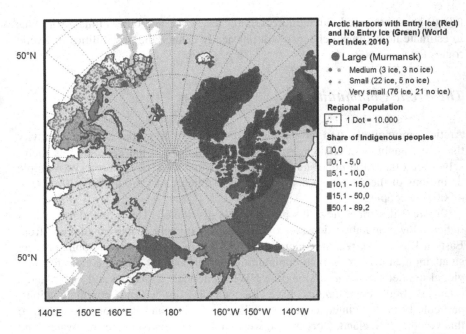

Fig. 1 Demographics of Arctic regions (population and ports) (Sources: ArcticStat and World Port Index 2016)

The economic and environmental changes under way are, however, requiring new transformation of the region (Newton et al. (2016)). The next section of the chapter explores historical analogies that may shed light on how best to manage and support these transformations.

Historical Analogy for the Current Day: Nineteenth Century Western Railroad Expansion

Western U.S. rail expansion consisted of boom-bust activity whose advantages and pitfalls economic historians and economists are still trying to understand fully today. One of the most famous controversies of economic history is over the role that railroad expansion played in nineteenth Century U.S. economic growth (see Majewski 2016). Some have argued the railroads acted as a major engine of growth, expanding opportunities by "growing ahead of demand" and leading development. Others suggest that railroads were mainly built only where they would be almost immediately profitable through transportation of natural resources or existing products to larger markets. Which of these theories is correct matters in helping understand future Arctic economic developments; the disagreement also highlights the potential challenges for the Arctic. Climate changes will increase the roles of ports in overall community connectivity. This will occur as maritime

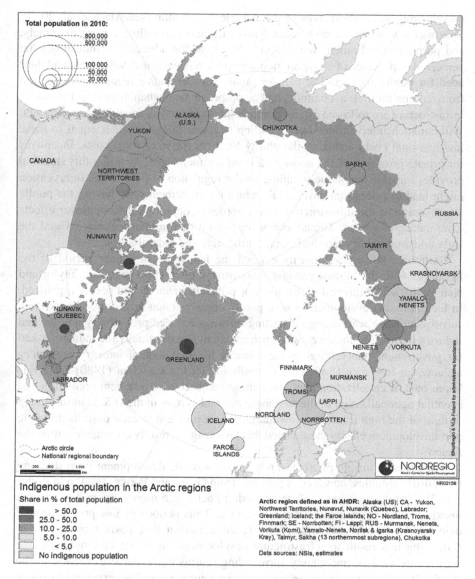

Fig. 2 Arctic population (Source: Nordregio at www.nordregio.se; image credit Johanna Roto)

trade and transport grow in response to less ice cover, while land-based transport is threatened by melting permafrost. Thus decisions about where to develop the ports further and how to incorporate them into networks will impact the future of the region directly.

The positive effects of national transport connectivity, e.g., through transcontinental railroads, were anticipated to be substantial even from earliest times. Advocates for federal support in assisting the development of these connections are present from the earliest days of the American Republic (see Atack et al. 2010). This was a time where both federal revenues and expenditures were extremely, and purposefully, limited (see Peacock 2004), so the advocacy demonstrates the perceived importance of transportation networks in American development and their need for public support. The call for public support and/or regulatory protection from competition is a common and appropriate one when high fixed costs of infrastructure are followed by low marginal costs of serving an additional customer. Without such interventions one is willing to lower price until it is equal to these low marginal costs, reducing the ability to cover (private) fixed costs. Incentives for private provision of the networked good are inefficient. Avoiding this situation provides impetus for public spending and/or regulation of ports. Such intervention might lead to prices that cover a fair return on investment or to having the public absorb some of the infrastructure cost in order to obtain positive spillover effects from the development. The advocacy for federal funding particularly assisted the early construction of roads and canals at the state and federal levels.

Railroads, however, primarily entered the landscape in a later period of the Republic, with the biggest expansions occurring after 1850 (see, e.g., Taylor and Neu 1956). Much railroad expansion was privately financed and poorly organized. In fact, railroad development was so poorly organized that series of collusions and subsequent price wars along competing railways reduced profitability and created industry losses. The losses grew to such an extent that eventually the industry asked for government intervention in the form of the first federal industry regulatory agency, the Interstate Commerce Commission (ICC); see Ulen (1980). That this boom-bust of cartelization and competition resulted in government regulation and eventual nationalization of the passenger railroad system in the U.S. is unsurprising in light of the high fixed cost, low marginal cost problem already described. Arctic port development efforts should heed this lesson and carefully coordinate within the region.

The research insight of greatest relevance for Arctic development from studying the railroad troubles, however, is possibly from Green and Porter (1984) and Porter (1983). Cartelization, where firms collude to act like a monopoly, collapsed with unexpected drops in demand for rail service. Thus periods of low prices through demand drops were exacerbated by failed collusion that pushed prices lower. Extending this result to infrastructure developments in Arctic communities, one sees highly volatile uncertainty surrounding returns on resource extraction in the Arctic. Port investments dependent on these returns should be expected to suffer similarly from the toll the volatility will take on profitability. As ports suffer from downturns in extractive industries, the intensity of competition between the ports is also likely to increase. This in turn may reduce profitability further. The railroad analogy strengthens understanding of how and why government cooperation across regions and ports will be vital to dampening the costly cyclical patterns of port activity in the future.

Recent research that attempts to reconcile the debate over railroads' contributions to growth suggests that when a community gained rail access to the national network during the late Antebellum Period (1850–1860), this had only a small positive impact on population density. On the other hand, it had a much larger impact on urbanization. Access to faraway markets concentrated the population by incentivizing it to move closer to depots (see Atack et al. 2010). The railroads served to concentrate rather than expand population. If such an impact extends to current Arctic communities, as early signs suggest, the effects on the economic and social conditions and outcomes of the communities will be significant. Communities closer to well-developed and integrated ports will continue to grow and become more globally market-oriented at the expense of more remote communities. Coordinated investments not only in ports and resource extraction, but also in community building, must evolve to avoid disastrous correlated losses like those the port of Churchill, Manitoba is currently facing; see the discussion below.

The ways in which rail networks chose where to locate in relation to the scarcely populated frontier reflected a mix of influences that relied on expectations of a highly uncertain future. This is similar to the uncertain future that Arctic communities face today. In many cases, rail investment, with its high fixed capital costs of development, resulted in financial ruin as competing lines and lack of coordination meant overcapacity and pricing wars.

Social costs were also high. Imported labor, primarily from Asia, was needed when local populations were insufficient. This has required long periods of social adjustment, and created a period of discriminatory history so unpleasant that it is frequently washed from the historical record altogether (see Takaki 2012). Similar problems appear to be growing both in practice (see, e.g., Amnesty International 2016 for evidence of correlations between violence toward indigenous women and resource extraction) and in theory (see, e.g., discussions of proposed Greenland mining; see Hansen et al. 2016; Nuttall 2013). Port infrastructure that is highly integrated with the needs of the local community presents a strong option for moving forward in the least cost manner to society.

In any case, one thing is certain: being "off the railroad path" was a difficult position for communities wishing to be included and prosper; this is also true in the Arctic today. Thus, it is important that policy making considers these complex network issues pertaining to the substitutability and complementarity of Arctic port development. In particular, the high investment costs for Arctic communities' port developments mean that financing assistance is likely to be an important factor in the ability of the scarcely populated indigenous regions to maintain connectivity in the face of reduced ice coverage. This assistance should be coupled with greater integration of governance covering the range of effects that will ensue as market forces and urbanization increase. Financing alone is insufficient as competitive market forces are likely to generate boom-bust cycles rather than balanced growth that considers full social costs.

Rail and Port Linkages: The Case of Churchill, Manitoba

The Port of Churchill, Manitoba on Hudson Bay is Canada's only Arctic deep-water port. Conceived and granted charters in the late nineteenth Century (1880), it was opened in 1931 and intended to stake Canada's claim in the Arctic as a strategic gateway to Europe, with the first grain shipments starting at that time (Port of Churchill 2016).

Analysis of railroad development in the nineteenth Century again provides insight. Canadian railroadization came slightly after U.S. railroad expansion; the Canadians thought they had learned how to avoid the boom-bust outcomes of the U.S. The Canadian government at the time played a more direct role in planning and financing than had the U.S. government in its railroad development, and initial expectations for profit were high. Nevertheless, Canadian railroad expansion resulted in financial distress and eventually nationalization in 1917. Subsequent research has shown that part of the failure could have been avoided if some of the investment in the Grand Trunk Railroad had been redirected to the Great Western Railroad instead (Carlos and Lewis 1992). As is frequently the case in high fixed cost (and in this case only seasonally accessible) investments, we might safely assume that the 50 year delay between the initial charters (made in 1880 during the frenzy for railroads) and the actual implementation of the port project at Churchill, completed in 1931, highlights the marginal financial returns to be expected on the route. This is because funds will first go to the projects with the highest expected returns. Projects with more uncertain, lower expected returns or that invite controversy over external costs will be delayed. As social objectives transition over time, delays may become even more prolonged. This occurs as new tradeoffs in land use and development become apparent; see Cain and Kaiser (2016). The Churchill port and connecting infrastructure have always been for social gains rather than economic ones.

The port serviced the prairies' wheat production and did lower transportation costs from the Canadian Western Prairies. Canadian wheat is in general at a significant global disadvantage due to high transportation costs. For this reason Canada created the Canadian Wheat Board to orchestrate the development of supporting infrastructure. Decentralization of the Board's activities, in combination with the industrial concentration in grain trade (there are five main companies) and rail transport (there are two companies) has significantly reduced Churchill's already limited competitiveness in recent years (Larsen 2016).

The Port of Churchill was privatized in 1997 and is currently owned by Omnitrax, a Denver-based U.S. company who also owns the railroad to the port. The company unexpectedly closed the port in 2016. The town has a population of 800 people, with 10% of the population working seasonally at the port (see Thompson Citizen 2016). Its mayor estimates that the port accounts for about 30% of the town's tourism revenues and 60% of the direct economic activity altogether, due to the historical average shipments of about 500,000 tonnes of grain shipped through the port. This figure has decreased in recent years, with only 280,000 tonnes shipped in 2015. The loss in shipping is directly related to the closure of the Canadian Wheat Board. This closure means that there is less direct governance of the industry. This has increased freedom of movement of goods to cheaper ports (see CBS News 2016).

The loss of Churchill as a functioning port is not expected to noticeably affect world trade in grain, either in price or quantity. It will however dramatically affect the local economic landscape. Though the port only operated seasonally, this seasonality fit with the timing of wheat harvests to render the port competitive. It could provide jobs and regional cash-based economic activity. The closure is in line with short term economic decisions for profit that often follow privatization and neglect the social and historical components of well-being. As Arctic port development goes forward in this global climate for economic profitability, northern communities will need to consider public investment goals broadly to carefully assess the best ways in which to promote long run social welfare. Connectivity generated by ports and transport infrastructure are vital components of economic security and trade. Public investment in port infrastructure must be supported.

Past and Present Use of the Arctic

Prior to 1969, a ship would complete a trip through the NWP on average once every ten years, with the rate beginning to increase in the 1950s (MacFarlane 2012). In 2012 alone, 30 known ships transited through the NWP and 17 ships are known to have done so in 2014 (Headland 2014). Yet, the Arctic is not a frontier. It has been used by Inuit for trade and communications over at least a thousand years. Its cultural heritage is part of the reasoning keeping people in the North. More recently, but still with long histories, are centuries of resource extraction. Examples on land include Spitsbergen (Svalbard) mining and the North American and Siberian fur trades, and at sea, whaling and walrus ivory. Boom and bust resource extraction for everything from fur seals and whales to coal, oil, and minerals throughout the Arctic have transformed local economies and the environments upon which they depend. This has had both positive and negative consequences. Trade and broader global contact have increased goods, information, and technologies available in the Arctic. Some traded goods, including alcohol and tobacco, have had dramatic negative consequences for indigenous populations in particular. Other negative consequences stem from resource depletion and ecosystem damages that affect the productivity of the landscape and the lives of those who depend upon it. How, then, will these economies continue to evolve?

In places like Svalbard, there have been no local indigenous populations to consider. In much of the Arctic, however, the extraction renewable and non-renewable resources has directly impacted traditional food supply, food security, and options for substitution from items; see Kaiser and Parchomenko, Chap. 6, this volume, and Bockstoce (1986). Without well-provisioned and organized ports, year-round access, or thoroughly considered governance of marine activities, transitions from indigenous hunter-gatherer communities to new forms of economic activity have been both slow and uneven as they are at the mercy of these boom-bust cycles. In the following sections, we examine past and present uses of Arctic port facilities to bring out trends and concerns for the broader social costs that will likely accompany economic development in the region. We address resource extraction, tourism, and fishing. Cargo shipping is treated within the context of resource extraction. This is due to the current relative importance of destination shipping as well as to the expectation that the TPR might soon altogether outcompete either shore-based route for the purpose of trans-Arctic shipping.

Destination Shipping and Resource Extraction

In this section, we analyze cargo and its impact on Arctic societies. Inbound cargo supplies remote Arctic communities with the many commercial goods that cannot be produced locally. These include any building and construction materials for consumer or industrial use. Outbound cargo carries resources extracted from communities to processing facilities and end-users.

Cargo

Trans-Arctic shipping of cargo through the NSR or the NWP is unlikely to become a substantial portion of economic activity (see Buixadé Farré et al. 2014). On the other hand, destination shipping, or shipping goods to and from locations in the Arctic itself, is likely to be a significant part of the Arctic ports' future. Cargo transport in the NSR is not a new phenomenon (see Fig. 3 and Pavlov and Selin 2015). Resource booms and busts in conjunction with political impetus to develop the Russian Arctic in the early twenty Century, followed by the collapse of the unifying political regime of the USSR, mean that the peak of NSR cargo shipping actually occurred in 1987, when 6, 579 thousand tonnes of cargo transited the NSR (Pavlov and Selin 2015).

In contrast, cargo shipping, and data on cargo shipping, in Arctic Alaskan and Canadian ports are virtually non-existent, see Table 1. As discussed above,

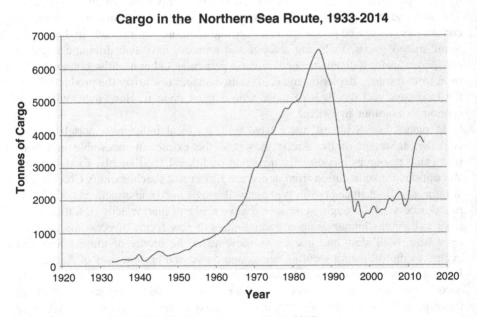

Fig. 3 Cargo in the NSR, 1933–2014 (See Pavlov and Selin 2015)

Table 1 Cargo table for Canadian ports (Source Statistics Canada)

Province	Port	Total tonnage handled (1000 t)				
		2007	2008	2009	2010	2011
Manitoba	Churchill	539.9	439.6	N/A	656.3	509.9
Nunavut	Chesterfield Inlet/Igluligaarjuk	5.0	17.9	N/A	28.1	96.7
Nunavut	Frobisher Bay/Iqaluit	N/A	N/A	39.7	38.5	41.3
Nunavut	Pangnirtung/Pannirtuuq	N/A	N/A	N/A	N/A	0.0
Nunavut	Nanisivik	N/A	N/A	N/A	2.6	4.4
Nunavut	Roberts Bay	N/A	N/A	N/A	18.6	0.0

the port of Churchill, Manitoba, was closed in 2016. It was the only deep-water (and yet seasonal) port in the Canadian Arctic. It handled a little over 500,000 tonnes/year. A few ports in Nunavut do have a sprinkling of data on cargo movement in the last decade. Still, for the ports that are identified and tracked by Statistics Canada, data is often not available and volumes are low, even for destination shipping.

Changes in climate and technology are lowering the costs and related barriers of access. This is once again increasing interest in Arctic activities. Consequential changes in technology range from ship characteristics like fuel types and ice-breaking capabilities to information processing like ice detectability and weather forecasting. Local and global stakeholders in Arctic activities can highly benefit from greater understanding of the interconnected socio-ecological systems and the needs and goals of the communities responding to these changes to avoid yet another boom-bust cycle. Shifts in perspective and opportunity are requiring rethinking of the infrastructure and technology demanded in Arctic communities, particularly ports, along with the goals of their use (see Kaiser et al. 2016).

Impacts on Societies

Resource extraction is likely to remain one of the highest sources of revenue and activity for many ports in the Arctic due to increasing (though volatile) global demands for resources. This perceived profitability is increasing public and private investment in port infrastructure. The new and prospering (or soon expecting to be prospering) ports of Novy, Sabetta, and Varandey in the Russian Arctic are all fully or partially owned by private hydrocarbon interests in cooperation with the Russian government (Staalesen 2017). At the same time, boom-bust cycles of resource extraction are likely accompanied by greater socially negative than positive impacts. The potential for these negative impacts in the North is high. Resource extraction in remote, inhospitable regions tends to impose high costs on indigenous populations. These costs may result in part from direct damages to wildlife and ecosystems that provide food supply. The costs are exacerbated when the indigenous populations are supplanted by imported labor rather than trained to take high paying positions in related industries. This creates inequality and other social consequences. An additional serious, but often overlooked, concern is that increased gender disparities should be expected to occur. A new report from Amnesty International (2016) on the increased violence against indigenous women in Fort John, British Columbia, for example, illustrates significant negative effects on populations.

Arctic Tourism

One option that is frequently presented for the continued viability of traditional resource-based communities that are trying to evolve within an encroaching outside world is *sustainable tourism*. Certainly, the month-long voyage of the Crystal Serenity cruise ship in 2016 through the NWP serves as a precursor of potential growth for future tourism. Is it more attractive to develop port infrastructure for the demands of coastal and/or trans-Arctic marine tourism than resource extraction? What would growth in this Arctic tourism entail? How would such tourism influence Northern communities? What is the role of Arctic port infrastructure, present and future, in determining these options?

In order to investigate these issues, we analyze the marine tourism (cruise) shipping demand and assess which companies already have plans for accessing the Arctic. Pashkevich et al. (2015) provide a recent overview of pan-Arctic tourism that highlights the fragmentation of governance and the concerns and hopes of many in the region. Demand for Arctic tourism might generate actions and policies that result in any of four possible states: *Arctic Race*, *Polar Preserve*, *Polar Lows*, or *Arctic Saga*. Underlying Arctic tourism is the reality that it should also be treated as a source of resource use and depletion. Tourism will have an uncertain net effect through the combination of market-measured economic growth with less quantifiable external economic and ecological damages. The role of governance in determining which state prevails, and whether harmonization is possible for the entire circumpolar region, is paramount. This governance should extend well beyond many existing definitions of tourism governance to include balanced growth in the use of underlying natural resources, labor and capital capacities of the region.

Arctic Race

The *Arctic Race* state is one with high demand for resource use with unstable governance that fosters competition between locations. It is likely to result in over-capitalization, increased risk, and increased damages. An example of such a situation might be tourism based on wildlife watching where regulations across locations differ significantly in either content or enforcement. If tourists vote with their feet based on the wildlife experience promised, they will often choose the least regulated environment that provides the closest access to marine mammals and other wildlife and/or landscape viewing. This can result in significant harm to both humans and wildlife, such as the loss of life that ensued when a gray whale crashed into a tourist boat in Mexico in 2015 (see Tuckman 2015).

Polar Preserve

The *Polar Preserve* state has low demand with stable governance. This generates slow growth where the Arctic functions not as a set of developing economies but instead as an eco-preserve. The Polar Preserve state may match global demands for use of the Arctic well, using the region primarily as a biological and ecological reserve with high value, particularly to wealthy nations and individuals such as comprise the eight Arctic countries. This, however, will do little to meet the needs of local and regional communities in need of economic transition due to the changing climate.

Polar Lows

The *Polar Lows* state combines low demand with unstable governance. This is a particularly poor combination that not only results in underdevelopment of the region, as with the Polar Preserve, but also foregoes the benefits of reduced development. This is because the lack of governance means that activities are poorly regulated, and are likely to impose significant damages on the community and environment.

Arctic Saga

The *Arctic Saga* state preserves Arctic ecosystem services for present and future use. The Arctic Saga state is expected to meet weak sustainability outcomes. These are outcomes where resources are used in a dynamically efficient manner to leave future generations no worse off than present ones, given the ability to substitute forms of natural, physical, and human capital for one another (Heal 2012). Furthermore, it may also meet stronger sustainability criteria. These stronger criteria recognize constraints in the substitutability amongst various forms of capital. This is particularly important in the development of Arctic transportation infrastructure due to the significant interrelations between cultural, ecological, and economic services in the North. Arctic communities are in many cases not developed market societies. Cash economy activities are viewed distinctly differently than other economic choices that remain embedded in cultural and ecological contexts. Policies that enhance the likelihood of the *Arctic Saga* scenario should be pursued as providing a balanced growth path of development.

Maritime Governance and Investment Efforts Toward *Arctic Saga* Outcomes

In 2008, the Ilulissat Declaration was adopted by the ministers of the five coastal states of the Arctic Ocean, i.e., Denmark/Greenland, Canada, Norway, Russia, and the U.S. (see Ministry of Foreign Affairs Denmark et al. 2011). Its aim was to send a strong political signal that these five states will act responsibly and in a collaborative manner with respect to the future developments in the marine Arctic. In general, maritime cooperation takes place within the framework of the eight nation Arctic Council (the five coastal states plus Sweden, Finland and Iceland) and the UN's *International Maritime Organization* (IMO), as well as through daily bilateral operations on *Search and Rescue* (SAR), environmental protection, and navigational safety.

The Arctic Council's few successes regarding pan-Arctic cooperation do help the region's governance. Two treaties with high relevance to marine tourism are the most significant accomplishments of the political organization in its 20 years of existence. The first, known as the 2011 Nuuk Declaration, is on the cooperation for Aeronautical and Maritime SAR (see Arctic Council 2011). The second, known as the 2013 Kiruna Declaration, is in Marine Oil Pollution Preparedness and Response (see Arctic Council 2013). Cooperation in governance that will assist in achieving an "Arctic Saga" outcome includes but is not limited to agreements on strict efforts toward wildlife protection, such as the U.S. Marine Mammal Act of 1972. This act requires vessels not to pursue marine mammals to within 50 yards of the animal. Subsequent higher voluntary standards in place, e.g., in parts of Alaska (see NOAA 2016) are also evolving. International agreements include the 1973 Agreement for the Conservation of Polar Bears made between the U.S., Russia, Canada, Denmark and Norway.

As the need for international agreements for polar bear protection might suggest, there are some particularly important safety and security issues for cruising in the Arctic. Cruise ships face incentives based on customer demands that increase the risks of accidents to both humans and wildlife. These include demands for visually appealing sites as well as marine wildlife. Arctic marine wildlife includes large, endangered marine mammals for whom human-animal interactions can easily result in injury or death. Floating ice, particularly icebergs, are also attractive to tourists (and polar bears), but risk damage to vessels. Furthermore, customers are willing to pay a premium for access to spots that other vessels may not be willing to risk visiting. Cruise ships therefore face a significant risk-reward tradeoff in designing cruise itineraries that other marine vessels may not.

Serious accidents already have occurred in recent years in Arctic waters. These accidents include the August 2016 sinking of a vessel off Ilulissat, Greenland carrying 23 cruise passengers and three crew members. This vessel was a private, local one that was serving as a tender, bringing passengers from a luxury cruise ship to shore. Several other local vessels were providing the same service, and were at hand for a quick rescue of all involved. The sunk vessel was loaded above its 22 passenger capacity (The Arctic Journal 2016).

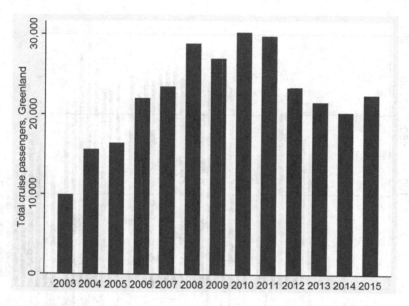

Fig. 4 Greenland cruise passengers, 2003–2015 (Source: bank.stat.gl)

The sort of informal arrangement described here is used frequently to accommodate larger cruise vessels and engage the entire community in earning extra income.
The risks are considerable, however, and port development must prepare more
effectively for loading and unloading passengers in small remote communities. The
number of Greenland cruise tourists to Ilulissat in 2015 was only 8, 250, which still
makes it the third-most visited port of call in the country (Statistics Greenland 2016).
Tourism numbers are still low, and infrastructure will need to grow rapidly if tourism
growth is to be successful and identifiable risks are to be mitigated. Figure 4 shows a
similar pattern to the Alaska cruise data (see Fig. 5). It further extends to more recent
years and shows that there was a downturn after 2011, for Greenland at least. This
2011 downturn was more significant than the effect of the financial crisis in 2009.
This downturn has been explained as "due primarily to external market conditions
beyond Greenland's control" (Mustafanezhad et al. 2016). In other words, the high
demand elasticity for cruise tourism due to the many substitutes for cruise and travel
activities can have significant impacts in small Arctic communities.

In 2015, Greenland acted to counter some of these forces by reducing taxes
affecting cruise pricing (Arctic Cluster of Raw Materials 2016). This is credited with
bringing up the 2015 tourism numbers and more gains are expected (VisitGreenland
2016). The tax shifts also favor larger vessels more than small vessels (VisitGreenland 2015). This could alter the risk factors either to the better or the worse. Larger
ships mean greater impact overall and lower levels of maneuverability, but they may
also allow greater safety and redundancy at sea.

For tourists, an Arctic cruise remains an expensive and less certain option,
particularly in terms of weather, than many other locations. Any Arctic cruise

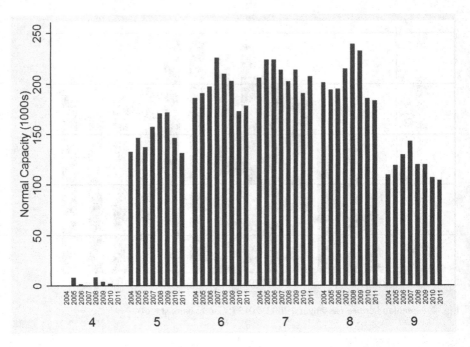

Fig. 5 Alaskan cruise capacity by month and year, 2004–2011 (Source: MARAD)

faces competition not only amongst Arctic locations, but also from other cruise destinations presented by the global cruise industry. The Greenland data attest this.

The data from Alaska also suggest that the industry is highly susceptible to economic downturns. There is little overcapacity in the Arctic as ships can be moved to alternative destinations in response to global demand shifts. Thus capacity and passenger nights track each other fairly well – voyages sell or are moved to another location. Figure 5 shows Alaskan cruise capacities by month from 2004 to 2011 (United States Department of Transportation 2017). Two patterns of interest emerge from the data. The first is the effect of the financial crisis creating a downturn after growth from 2004–2009. This downturn may have been further exacerbated by a set of new taxes levied on passengers and cruise ships from 2007–2010 (Resource Development Council 2017). This highlights the sensitivity of the industry to price fluctuations and unilateral cost-raising regulation for a destination. Thus straightforwardly prescribed economic tools such as a tax to recoup damages from external costs to, e.g., the environment may be ineffective in fostering community development. The high elasticity of demand makes such a tax a deterrent. No tax revenue is earned, plus the initial base of activity is reduced, if mobile capital departs for less costly locations.

The second pattern is the connection between the length of the season and the economic conditions. April, May, and September capacities have more room to expand and contract than June, July, and August. The summer months appear to have

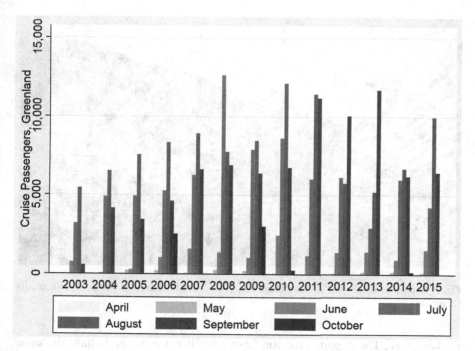

Fig. 6 Greenlandic cruise passengers by month and year, 2003–2015 (Source: bank.stat.gl)

a top capacity of about 225,000 passengers/month. Excess demand is then met with more shoulder season offerings. These shoulder season periods, particularly April and May, have greater weather risk and may result in lower consumer satisfaction. This will be especially true if, e.g., the timing of wildlife migrations is not in synchronization with the cruising. The weather risks also impact safety and security, increasing the likelihood of accidents.

For comparison, we also return to the Greenlandic cruise data and include the passenger days by month in Fig. 6. Here we see a similar expansion into the shoulder seasons. These are actually longer in Greenland than they are in Alaska, extending into October in some years. Greenland has also seen a small shift to August and September passengers over July.

Adding capacity in the high season brings its own problems. Port infrastructure is certainly one of these. Figure 7 shows the large cruise ships that serve as floating hotels and dominate the landscape. There may be up to five of these ships in the Ketchikan port at once. The cruise terminal itself is in large part a creation of the cruise line companies. The companies have created a tourist village staffed by seasonal workers selling trinkets from around the world. We provide data on seasonal employment in Ketchikan in year 2015 as an example, see Table 2.

The labor force increased by 27% for the July peak over February in 2015, with much of this increase coming from imported seasonal labor. This expansion requires infrastructure as well as governance that protects the rights of locals and visiting

Fig. 7 Cruise terminal, port of Ketchikan, AK (Photo Credit: Barek, 4 Jul 2008, Wikimedia Commons)

workers alike. The imported tourism labor force is not expected to have the same levels of consequence as extractive industries because the pay differential and the gender disparities are not as great. Planning of port infrastructure is still essential, however. In June, 2016, for example, a mildly windy approach to a berth on the Ketchikan docks resulted in a crash estimated to cost the facility $2–3 million dollars (see Shedlock 2016). Additional costs came from required repairs to the ship.

Currently, Russian Arctic marine tourism levels are very low. Efforts are under way to increase the numbers. These efforts, to some extent, reflect the sorts of port developments we suggest are best suited to the Arctic economy. These efforts include the establishment of, and investment in, the 1.5 million hectare Russian Arctic National Park encompassing parts of the Kara and Barents sea around Novaya Zemlya. Approximately 60% of the park territory is marine (see Pashkevich and Stjernström 2014).

The efforts also highlight failures in Russian and Pan-Arctic governance. Corrections of these failures could reduce impediments to the cruise industry. Pashkevich and Stjernström (2014) cite these as the "unpredictable nature of state control influencing tourism destination development, lack of coordination among the stakeholders on all levels of governance, [and the] overall low level of expertise in hospitality and tourism management." The role of the state in Russian Arctic development has not facilitated multi-use activities. Many nature preserves that have come into place in the last century do not allow for much human activity. In these locations, Russia is primarily on a path to the Polar Preserve outcome described above. Rather than utilizing synergies from Russian Arctic investment in, e.g., cargo and military transportation, Russian institutional settings keep much of the Arctic out-of-bounds either for foreigners or for all potential tourists, including Russians.

Table 2 2015 Labor force data Ketchikan, AK (Source: State of Alaska Department of Labor and Work Force Development (laborstats.alaska.gov))

	Jan	Feb	Mar	Apr	May	Jun	Jul	Aug	Sep	Oct	Nov	Dec	Annual
Labor force	6,410	6,397	6,474	6,687	7,313	7,697	8,124	7,946	7,530	6,702	6,471	6,528	7,023
Employment	5,851	5,809	5,922	6,199	6,881	7,227	7,720	7,561	7,137	6,236	5,977	6,034	6,546
Unemployment	559	588	552	488	432	470	404	385	393	466	494	494	477
Unemployment rate	8.70%	9.20%	8.50%	7.30%	5.90%	6.10%	5.00%	4.80%	5.20%	7.00%	7.60%	7.60%	6.80%

Thus despite the significantly higher levels of infrastructure in the Russian Arctic, tourism is far behind other areas in the region.

Small impositions are also costly. The number of cruise vessels that visited Franz Josef Land and Novaya Zemlya from 2000–2013 only totals 62 – an average of four or five per year. The high elasticity of demand lowers the ability to impose non-price tariffs as well as higher prices. Deterrents to more vessels include requirements that cruise ships accommodate, free of charge, three to five park officials and bear guards to serve as guides and protection. This requirement is followed because the bear guards are the only participants allowed to carry guns. The cost of the extra officials is increased due to the fact that they often have no tourism experience that enhances the passengers' enjoyment. Integrated tours with, e.g., Svalbard are made prohibitively costly by requirements to stop in Murmansk for visa clearance. These requirements also involve uncertainties and extra costs to secure. Both the overall cruise passage and every tourist on board must have their full itineraries cleared months in advance of the trip. This lengthy and inflexible process is due in part to the fact that all permits must be signed by the prime minister directly (Pashkevich and Stjernström 2014).

Fishing in the Arctic

Fishing in the Arctic is both commercial and local. Local fishing consists both of fishing tourism and subsistence fishing. There remains significant subsistence use in indigenous communities, particularly in areas of the NWP. Commercial use, particularly in the North Atlantic, Barents Sea, and Bering Sea, provides seafood to customers throughout the world. The Barents and Bering Seas also serve as entrances or exits to Central Arctic Ocean navigation. Fishing is also an important draw for Arctic tourism throughout the Arctic. Some locations, such as Iceland, currently have their fishing tourism more focused on their inland and shore-based fishing of anadromous species than coastal fishing. This is changing as excess capacity in the Icelandic coastal fleet combines with rapidly increasing tourism to draw tourists to the sea as well. Port infrastructure will need to respond accordingly, with access to shore-based tourists as well as cruise-ship based ones. As fishing tourism grows, recreational limits – and facilities at ports for monitoring these limits – are also likely to need increased development.

Subsistence consumption includes marine mammals. Fishing efforts led the push from southern areas into the Arctic centuries ago, in both the Pacific and the Atlantic. Vikings in Greenland followed the walrus in the Atlantic (Kintisch 2016); Asian fishermen followed fish and marine mammals north into the Bering Sea (McGhee 2017). Southern-based fishermen and Inuit alike had extensive knowledge about Arctic navigation from which many Arctic explorers failed to learn. As Barry Lopez elegantly puts it in his book *Arctic Dreams*, fishermen were "no doubt off the coast of Newfoundland before Cabot, in Frobisher Bay before Frobisher, in Hudson Strait before Hudson, and in Lancaster Sound before Ross arrived"

(Lopez 1999). Today, fishing is a less profitable pursuit in the Arctic than non-renewable resource extraction. Adages like fish are measured in pennies, while oil and gas are measured in dollars highlight the relative interest of the two industries to economic development efforts. This sort of expression clearly and erroneously ignores cultural and ecological values. Recently, even the basic economic arguments it relies upon have been shifting. An average (4.5 kg) Norwegian farmed salmon fetched a higher price than a barrel of Norwegian North Sea crude oil in January 2016 (see Berglund 2016). Given the multifaceted fishing interests in the Arctic, future port infrastructure in most cases must consider these combined subsistence, tourist, and commercial fishing interests.

Fishing vessel presence in Arctic waters from 2012 to the end of 2016 is revealed in Fig. 8. Panels 1–6 show voyages of vessels flagged to the five Arctic coastal countries (Denmark, Iceland, Norway, Russian Federation, United States) as well as Iceland. Panel 7 shows all vessel voyages with *Automatic Information System* (AIS) technology. The vessel tracks are the lightest shading on the maps; the more intense the light, the more vessels this represents. The data come from Global Fishing Watch (globalfishingwatch.org), one of a growing number of information outlets where fishermen and the public now have free information about vessel activities. AIS trackers are required by IMO SOLAS 19 regulations on all "vessels over 300 gross tonnage and upwards engaged on international voyages, cargo ships of 500 gross tonnage and upwards not engaged on international voyages and all passenger ships irrespective of size" (see IMO 2016). The requirement came into force at the end of 2004. With the data, one can follow individual vessels or correlated trends in fishing fleet behavior. This allows one to better understand and regulate the fisheries industry. For example, one could see if a port was being used heavily by vessels trying to participate in illegal, undocumented or unregulated (IUU) fishing. With this information in hand, regulators could position inspectors at ports, or otherwise adjust regulation and enforcement accordingly. The availability of the data lowers the cost of enforcement and increases the possibility of achieving improved growth outcomes that successfully conserve fisheries for long term sustainable use.

In this chapter, we use the agglomerated data to illustrate the density and locations of fishing vessels in the past four years for the different regions of the Arctic. From the first 6 panels of Fig. 8, we can see that Arctic flagged vessels, to a great extent, fish the sovereign waters of their *Exclusive Economic Zones* (EEZs). Comparison to the final panel, where vessels of all flags are shown, suggests that other vessels are using these fishing grounds as well. This is visible in that there are more areas that show fishing activity than in the individual panels showing only that nation's fleet. The increased vessel presence is particularly noticeable in the Barents Sea loophole and in the international waters surrounding Svalbard, along the Greenlandic coast, and to a lesser extent in the Chuckchi and Beaufort Seas north of Alaska and Canada. The overall intensity of use, evidenced by the density of voyages within the EEZs, does not appear to be much higher in total than for each country in its own waters. Thus EEZs in the north appear to be effective in limiting entry to own-state-flagged vessels. This is in accordance with the generally strong governance of these states in protecting their domestic fishing interests.

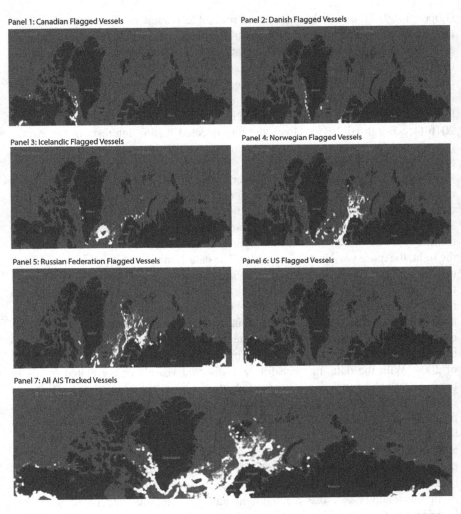

Fig. 8 AIS tracks of fishing vessel traffic in the north, 2012–2016 (Source: globalfishing-watch.org)

The images in Fig. 8 highlight the current absence of significant fishing activity in the northernmost waters. The five coastal Arctic states agreed to a moratorium on fishing in the Central Arctic Ocean in summer 2015 (Canada et al. 2015). The countries took this step to try to preempt actions by fishermen interested in opening up new fishing grounds as the costs of access decrease. There is also the expectation that existing fishery species ranges will expand northward. This is already happening for some species and locations, including for northern shrimp (*Pandalus borealis*) in Greenland. These shifts will include new demands on port infrastructure. For instance, the case of the introduction of the Red King Crab into the Barents Sea highlights how demands for Arctic port infrastructure are likely to change with climate change impacts and related invasion of species; these crabs are most profitable if they can be exported live, which requires rapid and thorough connectivity from crab pot to a landing and processing facility, and then on to ground and air transportation to markets.

The commercial value of the fisheries is not the only consideration. Small vessel coastal fishing fleets, such as the Norwegian fleet, are not just a way to provide tax revenues to the state. These fleets form the fabric of local communities in the Arctic. Changes in Arctic climate and species presence are transforming opportunities for these fishermen and their communities.

In the case of the Red King Crab, the Russians and the Norwegians are taking two different approaches. The Russians are fishing the species with large vessels capable of on-board processing, bypassing shore infrastructure needs to a large extent. The Norwegians, on the other hand, want to gain high marginal value on the crab through live year-round export incentivized on world markets. They see the crab as a way of supporting the declining northern coastal fishing fleet. The long decline has slowed with a recent tapering off in Finnmark, the northernmost Norwegian county and the only one whose residents are allowed to fish for Red King Crab. This stemming of the decline is not occurring in the next most northern counties, Troms and Nordland. This is illustrated by their continuing decline in vessel registrations in Fig. 9.

To date, the crab is confined mainly to Eastern Finnmark, a situation which the Norwegian authorities would like to maintain. This is because it has determined the crab to be an invasive species that should be contained from spreading (see Sundet

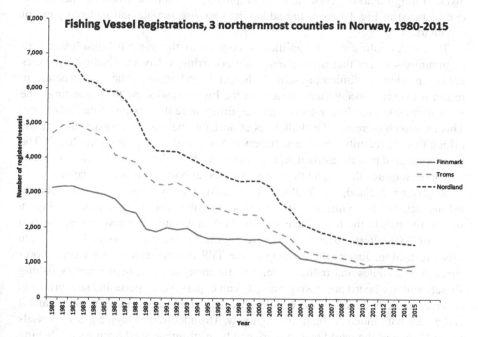

Fig. 9 Vessels registered in Northern Norway (Source: Norwegian Maritime Authority)

Fig. 10 Red King Crab landing facility in Trollbukt, Laksefjord (Photo: Brooks Kaiser, 2016)

and Hoel 2016). The abundance of the crab is creating a boon for the fishermen, but the remoteness of Finnmark and the special demands of live crab have meant that shore facilities for handling live crab must be located very near to the road and air transport, *and* to the crab pots, in order to minimize total travel times to distant foreign markets. New facilities employing 4–5 on-site workers, such as the one pictured in Fig. 10 are being added in otherwise remote and isolated parts of Finnmark.

This decentralization of facilities is counter to the general consolidation of community services that is simultaneously occurring in Norway. National policies are downsizing or eliminating schools, hospitals and other public infrastructure in return for more concentrated services in the bigger towns. At the same time, the most profitable maritime activities are requiring more dispersion of the fishermen. This mismatch increases the challenges of decision-making for investments and the difficulties for recruitment of new fishers to the industry (see Sønvisen 2013). The development of port infrastructure, even in the most developed and wealthy of Arctic locations, requires thorough integration with social policy if it is to succeed.

Regarding Iceland, Fig. 8 Panel 3 (Iceland) clearly shows Icelandic coastal fishing activity is significant. It is well known that commercial fishing plays a substantial part in the Icelandic economy. In Arctic waters, however, there are two types of commercial fishing. These are fishing by coastal fleets and large scale offshore fishing. Just as the opening of the TPR can be seen as bypassing shore-based Arctic routes and reducing demand for their services, high capacity fishing vessels with on-board processing capacity can bypass Arctic ports and take products to larger markets directly. This was the case with nineteenth century whaling, and it is the case with much northern fishing today. The adoption of high capacity vessels around much of the world was exacerbated by well-intentioned open access fishing regulations. Examples include seasonal restrictions that promoted capitalization of the fleet in order to harvest as many fish as quickly as possible.

Since the failures of these open access regulations have become known, *Individually tradeable quotas* (ITQs) are now presented as a preferable regulatory method. This permit system does manage the stock of fish in a more efficient manner with less overcapitalization of the fleet. The effect on ports, however, is unclear, and there is no better example of this than Iceland. Iceland moved to ITQs in the mid 1980s and experienced a rapid decrease in the number of fishermen holding quota. Concentration of quota into a few hands also increased. This has generated increasing inequality amongst remaining fishermen, turning formerly independent fishermen into laborers for others. Shore-based jobs in the ports decreased by a third – from about 10, 500 to 7, 000 employees – as processing moved to on-board operations (Olsen 2011). Net outcomes have been contentious, since distributional impacts have been so diverse (Eythórsson 2000).

Cross-Industry Considerations

In this chapter, we consider two important impacts that affect all countries and industries operating in the Arctic. These are seasonal constraints and the role of information.

Seasonalities: Development Impacts

As should be apparent, climate – particularly, but not only, sea ice – has played an important role in shaping transformations of Arctic communities. Nineteenth Century marine activities in the Arctic focused on harvests of whales and other food sources. Demands for infrastructure were limited since the whaling and walrusing ships generally functioned as self-contained offshore factories. Even so, the southerners' whaling fleets wreaked havoc on local communities' sustainability, and the boom in whaling created whale, walrus, and then Inuit population busts throughout the Inuit Arctic (Bockstoce 1986). Winter survival became more difficult with lower food supplies, and no newcomers remained in the area to trade imported goods or technologies to assist survival and transitions; they departed with their ships in the fall before ice destroyed their ship-factories. Summer and early fall were the "open" seasons for North-South communication by ship in much of the Arctic. These seasons are, however, to a large extent the "closed" seasons for trade and transport amongst indigenous communities (circumpolar communication). This is because they use the sea ice for hunting and transportation amongst small scattered communities. Thus simultaneous seasonal resource extraction and the continuation of traditional indigenous lifestyles could and did coincide to a greater degree than in many other developing regions, despite devastating population impacts (see

Bockstoce 1986). When resource extraction switched from those that had important competing uses as subsistence to non-renewables that had little local use, these population pressures on whales, walrus and men eased up to some extent.

Today, the resources have shifted to oil, gas, minerals, or tourism services, but overall extraction pressures and impacts on indigenous lifestyles are rising again. When looking at data from NORDREG, an increase in traffic through the Northeast and NWP is evident; see Environment and Natural Resources (2015). Furthermore, Arctic ports stand to play an increasingly important role in communications with the rest of the world as land-based travel is expected to deteriorate significantly. This deterioration is primarily due to melting permafrost; see, e.g., Stephenson et al. (2011). The magnitude of this loss ranges from a reduction of 11% in winter-road-accessible land area in Greenland (where road travel is already scarce) to 82% in Iceland, with an overall loss in Arctic regions of 14%. The expected gain in maritime access totals 23%. This comes mostly from increased access to the high seas (a 406% increase). Still, Canada, Greenland and the Russian Federation all are expected to gain over 15% in area. Iceland, Norway, and the U.S. on the other hand stand to gain little or no additional area (Stephenson et al. 2011).

The loss of land transport options due to melting ice roads and the gain of marine traffic zones are seasonal. For instance, during July-September, the Arctic is projected to be open for maritime shipping in the NWP, NSR, and TPR routes (Stephenson et al. 2011). Studies conducted by Stephenson et al. (2011) suggest that a substantial increase in maritime traffic will occur for Canada, Greenland, Russia, and the United States, although ships will still require their own ice-breaking capacity or ice-breaking services from another vessel.

The Role of Information

Information about environmental conditions, ship locations, and their activity is rapidly becoming more and more available to shipowners, government regulators, and the public around the world, including the Arctic. The rate of change of use for remote sensing tools is formidable. Live and historical AIS data is available from sites such as *marinetraffic.com* and *vesselfinder.com*, with live data freely available and historical data available at fairly low cost. This reduction in the costs of information can be expected to have significant positive effects on safety and security, as intended by the regulations. There are larger consequences as well. Some remote sensing products combine AIS information with self-reporting and other remote-sensing inputs such as weather. An example of this is BarentsWatch (*barentswatch.no/en*). BarentsWatch is primarily a tool to aid the fishing industry in the Barents sea. On the website one can easily identify fishing locations in use and what type of fishing is occurring through differentiated markers. New species are moving north at record rates. Fishermen will be some of the first and most influential observers of these shifts. Thus tools like BarentsWatch will serve to inform the fishermen and those who process the fish and make shore investments based on predictions regarding fish stocks. These tools can also inform about their need for regulation and/or compliance with that regulation.

Still, the Arctic region is lacking in much basic information about its economies and prospects. There is significant uncertainty surrounding the biggest hope of the past decades – oil and gas reserves. This uncertainty only partly stems from world forces creating highly volatile prices. It also comes from the challenges of accurately identifying new resources. The U.S. Department of the Interior public estimates of reserves of both oil and gas are quite large (U.S. Department of the Interior 2008). The subsequent activity in, e.g., the Beaufort (Shell) and off the Greenland coast (Cairns) has, however, resulted in disappointment. This disappointment has been increased by lower global prices and profit expectations. The departure of Shell from the Arctic proved a devastating blow for port development aims in Northern Alaska.

Conclusions

This chapter describes past and current conditions in the Arctic as they pertain to port infrastructure and use. The chapter looks at the resource extraction, tourism, and fishing industries in particular. In so doing, we find that economic incentives in northern communities have tendencies that generate cyclical economic outcomes that require governance to mitigate if balanced growth is to develop for the region. Trans-Arctic shipping is not investigated in large part because it is expected to provide only short term opportunities for local communities at best, due to the lower-cost potential of the TPR that can bypass the shore-based NWP and NSR routes.

Of these tendencies, we find that resource extraction has the most risk of creating negative social costs for communities. These social costs range from stranded investments to gender-based violence and inequality. Price fluctuations outside of the control of Arctic communities dictate demand for the resources and leave the communities vulnerable to boom-bust cycles. Skills needed in the industry do not match skills available in the local populations, so job creation for local communities is expected to be low. At the same time, outside labor, sometimes with significant cultural differences, is arriving or anticipated to arrive. This brings additional social conflicts and challenges as market-based activity grows and favors high wage earners over local subsistence on the marine environment. The profitability of oil and gas resources in Russia has invited private investment in several ports. This will serve the industry, but diversified communities cannot easily be built on such foundations, and local stakeholders stand to be disenfranchised.

Arctic fisheries, which operate on four intertwined levels of subsistence, recreational, coastal commercial and offshore commercial, have long histories throughout the area. Most are currently well regulated. Climate changes are affecting the stability of these regulations and outcomes, however. The introduction of new species and the movement of existing commercial species into new areas are shifting the port infrastructure needs. There have been recent positive and negative developments. The privatization of Icelandic fisheries through ITQs has increased overall Icelandic wealth, but left many fishing communities and their ports with few economic prospects as property rights move to larger vessels offshore and citizens move to bigger towns. In Norwegian waters, the introduction of the Red King Crab is creating ecological damages as well as economic benefits for coastal fishermen. The requirements of the live crab industry, however, are demanding a diffusion of facilities at a time when cost savings and economies of scale from government planners have been trying to move in the opposite direction, consolidating facilities.

Arctic marine tourism is profitable and growing, but demand elasticity for Arctic cruises is high as capacity can be rapidly adjusted by moving vessels. Demands by tourists for shopping and entertainment in ports are leading to imbalances in what the local population can provide (and staff), and what the cruise companies will do to supplement this capacity with, e.g., outside labor or internal training. This imbalance may strain local relations in similar ways to resource extraction if imported labor is seasonal and used to supplant local populations rather than support them. As cruise vessels become larger to accommodate more passengers, port facilities must invest in capacity that will only be used for a few visits a year, or use higher risk methods for bringing passengers to shore in order to avoid being bypassed altogether.

The challenges for Arctic port development are significant. The choices made over both port infrastructure and industry regulation will have long-ranging impacts on the available development paths in the region. Past experiences with high-fixed-cost, low-marginal-cost transportation networks like North American railroads provide cautionary lessons. These lessons suggest that public investment in port infrastructure can improve the likelihood of positive outcomes but that it should also be accompanied by regulatory oversight of the industries that develop along the transportation networks. This regulatory oversight should work to reduce over-capitalization. It should seek to promote social equality where resource extraction is prominent. It should also seek to enhance cooperation for the provision of networked tourism opportunities to support both of the aforementioned goals. It should conserve its living marine resource base of marine mammals and fisheries to meet a wide variety of interests and uses in the present and future. The realization of these goals will require cooperation within and across Arctic nations.

References

Aksenov, Y., Popova, E. E., Yool, A., Nurser, A. J. G., Williams, T. D., Bertino, L., & Bergh, J. (2016). On the future navigability of Arctic sea routes: High-resolution projections of the Arctic ocean and sea ice. *Marine Policy, 75*, 330–317.

Amnesty International. (2016). Canada: Out of sight, out of mind: Gender, indigenous rights and energy development in Northeast British Columbia. Internet Source: https://www.amnesty.org/en/documents/amr20/4872/2016/en/. Last call: 19 Jan 2017.

Aporta, C. (2009). The trail as home: Inuit and their pan-Arctic network of routes. *Human Ecology, 37*(2), 131–146.

Arctic Cluster of Raw Materials. (2016). Greenland and benchmarking report 2016. Internet
    Source: http://acrm.dk/?knowledge-center=greenland-benchmarking-report. Last call: 30 Mar
    2017.
Arctic Council. (2011). Nuuk declaration (2011). Internet Source: http://hdl.handle.net/11374/92.
    Last call: 19 Jan 2017.
Arctic Council. (2013). Kiruna declaration (2013). Internet Source: http://hdl.handle.net/11374/93.
    Last call: 19 Jan 2017.
Atack, J., Bateman, F., Haines, M., & Margo, R. (2010). Did railroads induce or follow economic
    growth? Urbanization and population growth in the American Midwest, 1850–1860. *Social
    Science History, 34*(2), 171–197.
Berglund, N. (2016). Salmon worth more than a barrel of oil. Internet Source: http://www.
    newsinenglish.no/2016/01/15/salmon-worth-more-than-oil/. Last call: 19 Jan 2017.
Bockstoce, J. R. (1986). Whales, ice, and men: The history of whaling in the Western Arctic.
    Seattle: University of Washington Press.
Buixadé Farré, A., Stephenson, S. R., Chen, L., Czub, M., Dai, Y., Demchev, D., Efimov, Y.,
    Graczyk, P., Grythe, H., Keil, K., Kivekas, N., Kumar, N., Liu, N., Matelenok, I., Myksvoll,
    M., O'Leary, D., Olsen, J., Pavithran, S. A. P., Petersen, E., Raspotnik, A., Ryzhov, I., Solski,
    J., Suo, L., Troein, C., Valeeva, V., van Rijckevorsel, J., & Wighting, J. (2014). Commercial
    Arctic shipping through the Northeast Passage: Routes, resources, governance, technology, and
    infrastructure. *Polar Geography, 37*(4), 298–324. http://www.enr.gov.nt.ca/state-environment/
    73-trends-shipping-northwest-passage-and-beaufort-sea.
Cain, L. P., & Kaiser, B. A. (2016). A century of environmental legislation. In S. Wolcott & C.
    Hanes (Ed.), *Research in economic history* (Chap. 1, Vol. 32, pp. 1–71. Bingley: Emerald
    Insight.
Canada, Denmark, Norway, Russian Federation, & United States of America. (2015). Dec-
    laration concerning the prevention of unregulated high seas fishing in the central Arctic
    Ocean. Internet Source: https://www.regjeringen.no/globalassets/departementene/ud/vedlegg/
    folkerett/declaration-on-arctic-fisheries-16-july-2015.pdf. The countries of Canada, Denmark,
    Norway, Russian Federation and the United States of America; Last call: 19 Jan 2017.
Carlos, A. M., & Lewis, F. (1992) The profitability of early Canadian railroads: Evidence from the
    grand trunk and great western railway companies. In *Strategic Factors in Nineteenth Century
    American Economic History: A Volume to Honor Robert W. Fogel* (NBER chapters, pp. 401–
    426). National Bureau of Economic Research, Inc.
CBS News. (2016). Churchill port closure could mean small shipment bump for Thunder Bay.
    Internet Source: http://www.cbc.ca/news/canada/thunder-bay/churchill-port-closure-thunder-
    bay-1.3697342. Last call: 16 Dec 2016.
Environment and Natural Resources. (2015). State of the environment: 7.3 Trends in shipping in
    the Northwest Passage and the Beaufort Sea. Internet Source: http://www.enr.gov.nt.ca/state-
    environment/73-trends-shipping-northwest-Passage-and-beaufort-sea. Last call 13 May 2016.
Eythórsson, E. (2000). A decade of ITQ-management in Icelandic fisheries: consolidation without
    consensus. *Marine Policy, 24*(6), 483–492.
Glomsröd, S., & Aslaksen, I. (2008). Presenting the economy of the North 2008. Internet Source:
    https://www.ssb.no/a/publikasjoner/pdf/sa112_en/sa112_en.pdf. Last call: 11 Nov 2016.
Green, E. J., & Porter, R. H. (1984). Noncooperative collusion under imperfect price information.
    Econometrica: Journal of the Econometric Society, 52(1), 87–100
Hansen, A. M., Vanclay, F., Croal, P., & Hurup Skjervedal, A.S. (2016). Managing the social
    impacts of the rapidly-expanding extractive industries in Greenland. *The Extractive Industries
    and Society, 3*(1), 25–33.
Headland, R. K. (2014). Transits of the Northwest Passage to end of the 2014 navigation season.
    Internet Source: http://www.americanpolar.org/wp-content/uploads/2014/10/NWP-2014-X-5-
    layout-for-PDF.pdf. Scott Polar Research Institute, University of Cambridge, UK, Last call: 03
    Apr 2017.
Heal, G. (2012). Reflections-defining and measuring sustainability. *Review of Environmental
    Economics and Policy, 6*(1), 147–163

IMO. (2016). AIS transponders. Internet Source: http://www.imo.org/en/OurWork/Safety/Navigation/Pages/AIS.aspx. Last call: 19 Jan 2017.

Intergovernmental Panel on Climate Change. (2001). *Climate change 2001: The scientific basis – a report of working group I of the intergovernmental panel on climate change.* Internet Source: http://www.grida.no/climate/ipcc_tar/wg1/pdf/WG1_TAR-Front.pdf. 2.2.5–2.2.6; Last call: 19 Jan 2017.

Johannessen, O. M., Shalina, E. V., & Miles, M. W. (1999). Satellite evidence for an Arctic sea ice cover in transformation. *Science, 286*(5446), 1937–1939.

Jørgensen-Dahl, A., & Wergeland, T. (2013). Shipping, resources, economic trends and alternative means of transport. In W. Østreng, K. M. Eger, B. Fløistad, A. Jørgensen-Dahl, L. Lothe, M. Mejlænder-Larsen, & T. Wergeland (Eds.). *Shipping in Arctic waters: A comparison of the Northeast, Northwest and Trans Polar Passages* (Chap. 4, pp. 83–145). Berlin/Heidelberg: Springer.

Kaiser, B. A., Fernandez, L. M., & Vestergaard, N. (2016). The future of the marine Arctic: Environmental and resource economic development issues. *The Polar Journal, 6*(1), 152–168.

Kintisch, E. (2016). The lost Norse. *Science, 354*(6313), 696–701.

Koenigk, T., Brodeau, L., Graversen, R. G., Karlsson, J., Svensson, G., Tjernström, M., Willen, U., & Wyser, K. (2013). Arctic climate change in 21st century CMIP5 simulations with EC-Earth. *Climate Dynamics, 40*(11–12), 2719–2743.

Larsen, L. (2016). An evaluation of the present situation for western Canadian grain farmers within a historical context. A report prepared for the Canadian Wheat Board Alliance. Internet Source: http://www.cwbafacts.ca/wp-content/uploads/2016/04/Evaluation-of-present-situation-for-Western-Canadian-grain-farmers_revised.pdf. Last call: 27 Mar 2017.

Lopez, B. (1999). Arctic dreams: Imagination and desire in a northern landscape. London: Vintage.

MacFarlane, J. M. (2012). Full transits of the Canadian Northwest Passage. Internet Source: Nauticapedia. http://nauticapedia.ca/Articles/NWP_Fulltransits.php. Last call: 03 Apr 2017.

Majewski, J. (2016). American railroads and the transformation of the antebellum economy classic reviews in economic history. Internet Source: http://www.eh.net/?s=American+railroads+and+the+transformation. Last call: 19 Jan 2017.

McGhee, R. (2017). The archaeological construction of aboriginality. In Hillerdal, C., Karlstrom, A., & Ojala, C. G., (Eds.). *Archaeologies of "Us" and "Them": Debating history, heritage and indigeneity.* London: Routledge.

Ministry of Foreign Affairs Denmark, Department of Foreign Affairs Greenland, and Ministry of Foreign Affairs Faroes. (2011). *Kingdom of Denmark strategy for the Arctic 2011–2020.* Internet Source: http://www.uniset.ca/microstates/mss-denmark_en.pdf. ISBN:561-5, Last call: 10 Dec 2016.

Morrison, D. (1991). The copper Inuit soapstone trade. *Arctic, 44*(3), 239–246

Mustafanezhad, M., Norum, R., Shelton, E. J., & Thompson-Carr, A., (Ed.). (2016). *Political ecology of tourism. Community, power and the environment.* London: Routledge.

Newton, R., Pfirman, S., Schlosser, P., Tremblay, B., Murray, M., & Pomerance, R. (2016). White Arctic vs. blue Arctic: A case study of diverging stakeholder responses to environmental change. *Earth's Future, 4*(8), 396–405.

NOAA. (2016). Marine mammal viewing guidelines and regulations Alaska regional office. Internet Source: https://alaskafisheries.noaa.gov/pr/mm-viewing-guide. Last call: 19 Jan 2017.

Nuttall, M. (2012). The circumpolar north: Locating the Arctic and sub-Arctic. In R. Fardon, O. Harris, T. H. J. Marchand, M. Nuttall, C. Shore, V. Strang, & R. A. Wilson (Eds.). *The SAGE handbook of social anthropology* (Chap. 1.2, pp. 270–285). London: SAGE Knowledge.

Nuttall, M. (2013). Zero-tolerance, uranium and Greenland's mining future. *The Polar Journal, 3*(2), 368–383.

Olsen, J. (2011). Understanding and contextualizing social impacts from the privatization of fisheries: an overview. *Ocean & Coastal Management, 54*(5), 353–363.

Pashkevich, A., & Stjernström, O. (2014). Making Russian Arctic accessible for tourists: analysis of the institutional barriers. *Polar Geography, 37*(2), 137–156.

Pashkevich, A., Dawson, J., & Stewart, E. J. (2015). Governance of expedition cruise ship tourism in the Arctic: A comparison of the Canadian and Russian Arctic. *Tourism in Marine Environments, 1*+(3–4), 225–240.

Pavlov, K. V., & Selin, V. S. (2015). Problems of development of freight traffic of the Northern Sea Route and methods of their decision. *Bulletin UGUES. Science, education, the economy. Series: Economy, 2*(12), 73–80.

Peacock, A. T. (2004). The growth of public expenditure. In Rowley, C. K., & Schneider, F. (Ed.). *The encyclopedia of public choice* (pp. 3–31). Springer. http://link.springer.com/book/10.1007 %2Fb108558.

Port of Churchill. (2016). History: Port of Churchill Hudson Bay Port Company. Internet Source: www.portofchurchill.ca/about/history. Last call: 16 Dec 2016.

Porter, R. H. (1983). A study of cartel stability: the joint executive committee, 1880–1886. *The Bell Journal of Economics, 14*(2), 301–314.

Resource Development Council. (2017). Alaska's tourism industry. Internet Source: http://www. akrdc.org/tourism. Last call: 24 Mar 2017.

Richter-Menge, J., Overland, J. E., & Mathis, J. T. (2016). Arctic report card 2016. Internet Source: http://www.arctic.noaa.gov/Report-Card. Last call: 30 Dec 2016.

Saxinger, G. (2016). Lured by oil and gas: Labour mobility, multi-locality and negotiating normality & extreme in the Russian far north. *The Extractive Industries and Society, 3*(1), 50–59.

Shedlock, J. (2016). Video: Cruise ship crashes into Ketchikan Dock. Internet Source: https:// www.adn.com/alaska-news/2016/06/03/video-cruise-ship-crashes-into-ketchikan-dock/. Alaska Dispatch News; Last call: 19 Jan 2017.

Sønvisen, S. A. (2013). Recruitment to the Norwegian fishing fleet: Storylines, paradoxes, and pragmatism in Norwegian fisheries and recruitment policy. *Maritime Studies, 12*(1), 1–8

Staalesen, A. (2017). Russian Arctic ports have best year ever. Internet Source: The Barents Observer. https://thebarentsobserver.com/en/industry-and-energy/2017/01/russian-arctic-ports-have-best-year-ever. Last call: 24 Mar 2017.

Statistics Greenland. (2016). Cruise ship statistics. Internet Source: http://www.stat.gl/dialog/main. asp?lang=en&sc=TU&version=201606. Last call: 24 Mar 2017.

Steinberg, P. E. (2015). *Contesting the Arctic – Politics and imaginaries in the circumpolar north* (International library of human geography). London: I.B. Tauris.

Stephenson, S. R., Smith, L. C., & Agnew, J. A. (2011). Divergent long-term trajectories of human access to the Arctic. *Nature Climate Change, 1*, 156–160.

Stuhl, A. (2016). *Unfreezing the Arctic*. Chicago: The University of Chicago Press.

Sundet, J. H., & Håkon Hoel, A. H. (2016). The Norwegian management of an introduced species: the Arctic red king crab fishery. *Marine Policy, 72*, 278–284.

Takaki, R. (2012). *Strangers from a different shore: A history of Asian Americans* (revised and updated edition). New York: Little, Brown and Company.

Taylor, G. R., & Neu, I. D. (1956). *The American Railroad Network, 1861–1890*. Urbana: University of Illinois Press.

The Arctic Journal. (2016). *Inuk and Sunk: Quick action prevented serious tour-boat accident.* Internet Source: http://arcticjournal.com/business/2512/quick-action-prevented-serious-tour-boat-accident. Last call: 19 Jan 2017.

Thompson Citizen. (2016). *Ashton asks government to stop catering to billionaires and save the port of Churchill.* Internet Source: http://www.thompsoncitizen.net/news/nickel-belt/ashton-asks-government-to-stop-catering-to-billionaires-and-save-the-port-of-churchill-1.3531622. Last call: 16 Dec 2016.

Tuckman, J. (2015). Tourist dies and two injured after whale crashes into sightseeing boat off Mexico. Internet Source: https://www.theguardian.com/world/2015/mar/12/jumping-whale-kills-canadian-woman. Last call: 19 Jan 2017.

Ulen, T. S. (1980). The market for regulation: The ICC from 1887 to 1920. *The American Economic Review, 70*(2), 306–310.

United States Department of Transportation. (2017). Maritime administration (MARAD) resources. Internet Source: https://www.marad.dot.gov/resources/data-statistics/. Last call: 24 Mar 2017.

U.S. Department of the Interior. (2008). Circum-Arctic resource appraisal: Estimates of undiscovered oil and gas north of the Arctic circle. Internet Source: USGS Fact Sheet 2008-3049: https://pubs.usgs.gov/fs/2008/3049/fs2008-3049.pdf. Last call: 04 Dec 2016.

VisitGreenland. (2015). Tourism report q3-q4 2015. Internet Source: http://tourismstat.gl/?lang=en. Last call: 30 Mar 2017.

VisitGreenland. (2016). Tourism strategy 2016–2019. Internet Source: http://corporate.greenland.com/en/about-visit-greenland/strategi-2016-2019/. Last call: 30 Mar 2017.

Conclusions

Brooks A. Kaiser, Melina Kourantidou, Niels Vestergaard, Linda Fernandez, and Joan Nymand Larsen

Abstract The ecosystem changes underway in the Arctic region are expected to have significant impacts on living resources in both the short and long run, and current actions and policies adopted over such resource governance will have serious and ultimately irreversible consequences in the near and long terms. The chapters in this book present a wide cross-section of research on Arctic Marine Resource Governance and its role in past and future regional development. They stem from a conference on this topic held in Reykjavik, Iceland in October of 2015. Several chapters delve into past, present, and future implications of fisheries resource use and management at multiple scales, complementing the breadth of focus. The complexities of the Arctic political, economic, and ecological environment mean that governance must accommodate multiple scales of use and concern. Rapid climate change – predicted to be more rapid and more influential in the Arctic than anywhere else on the planet – means that shifts in ecosystems and the resources they provide will require adaptive, ecosystem-based management to successfully navigate the uncertainty and change underway.

Keywords Arctic Fisheries • Arctic marine resource governance • Arctic economic development • Arctic Council • Sustainable Development in the Arctic

B.A. Kaiser • M. Kourantidou • N. Vestergaard (✉)
Department of Sociology, Environmental and Business Economics, University of Southern Denmark, Niels Bohrs Vej 9, 6700, Esbjerg, Denmark
e-mail: baka@sam.sdu.dk; nv@sam.sdu.dk

L. Fernandez
Department of Economics, Center for Environmental Studies, Virginia Commonwealth University, Snead Hall, 301 W. Main Street, 844000, Richmond, VA, 23284-4000, USA
e-mail: lmfernandez@vcu.edu

J.N. Larsen
Stefansson Arctic Institute, University of Akureyri, Borgir, Nordurslod, 600, Akureyri, Iceland
e-mail: jnl@unak.is

© Springer International Publishing AG 2018   219
N. Vestergaard et al. (eds.), *Arctic Marine Resource Governance and Development*,
Springer Polar Sciences, https://doi.org/10.1007/978-3-319-67365-3_10

The ecosystem changes underway in the Arctic region are expected to have significant impacts on living resources in both the short and long run, and current actions and policies adopted over such resource governance will have serious and ultimately irreversible consequences in the near and long terms. The chapters in this book present a wide cross-section of research on Arctic Marine Resource Governance and its role in past and future regional development. They stem from a conference on this topic held in Reykjavik, Iceland in October of 2015. Several chapters delve into past, present, and future implications of fisheries resource use and management at multiple scales, complementing the breadth of focus. The complexities of the Arctic political, economic, and ecological environment mean that governance must accommodate multiple scales of use and concern. Rapid climate change – predicted to be more rapid and more influential in the Arctic than anywhere else on the planet – means that shifts in ecosystems and the resources they provide will require adaptive, ecosystem-based management to successfully navigate the uncertainty and change underway.

The themes of the conference were:

1. Global management and institutions for Arctic marine resources
2. Resource stewards and users: local and indigenous co-management
3. Governance gaps in Arctic marine resource management
4. Multi-scale, ecosystem-based, Arctic marine resource management

Many of the chapters embrace aspects from multiple themes. The intersections across themes serve to highlight the Arctic's longstanding interconnectedness in global affairs in spite of its distance and the relative isolation of many of its peoples. The multiple scales of engagement with Arctic resources, from local indigenous subsistence users to multinational corporations and global warm-glow oriented or resource-hungry citizens, require that we refine some of the basic premises of resource management to incorporate better the competing demands and variations in understandings of value. The governance structures that emerge from negotiations over these competing demands will have very long run impacts on the welfare of northern communities. They will directly affect the direction of infrastructure investment, human and physical capital stock developments, resource flows, and the terms of trade between *in-situ* stakeholders of the Arctic's natural capital and the rest of the world.

The opening chapters focus on some of the important direct governance issues relating to the Arctic's opening to the rest of the world. Chapter "Regulating Fisheries in the Central Arctic Ocean: Much Ado About Nothing?" (Rayfuse) transitions the discussion of global concerns into the specifics of the fishery industry and the importance governance decisions have on present and future living marine resources in the Arctic.

The shifting structure of fisheries management in the Arctic neatly summarizes many of the challenges facing the Arctic due to its multiplicity of stakeholders in the past, present, and future. Current state-level management is rapidly evolving into co-management, led by Canada, as discussed in chapter "A Half Century in the Making: Governing Commercial Fisheries through Indigenous Marine Co–

management and the Torngat Joint Fisheries Board" (Snook et al.). Not all of the Arctic, especially the northernmost waters, is covered by international fishing agreements. The number of international agreements is increasing, however. Current international management is anchored in some locations by such longstanding agreements as the Joint Norwegian-Russian Fisheries Commission (est. 1974), the North Atlantic Fisheries Organization (est. 1979 from the International Commission of the Northwest Atlantic Fisheries, est. 1949), and the North East Atlantic Fisheries Commission (est. 1982). The current and proposed plans face the normal challenges of international cooperation but also new complexities stemming from climate change's known and unknown but anticipated potential impacts on fish populations and locations (Christiansen et al. 2014).

Chapter "Long Run Transitions in Resource-Based Inuit Communities" (Kaiser and Parchomenko) then segues from discussions of current models of bio-economic analysis and fisheries management to historical understanding of the evolution of some of the indigenous governance gaps that current governance is acting to remedy. Redressing the historical imbalance of bargaining power in initial resource trade flows from the Arctic to the rest of the world is a paramount concern for Arctic indigenous communities. This ties directly into questions of how far Arctic Council expansion to non-Arctic states should go, as increases in power to non-Arctic states is seen as a potential threat to the ability to maintain and improve societal well-being for local and indigenous communities (Knecht 2017).

The remaining chapters also draw on historical lessons to guide decisions about policy aimed at reducing invasive species introductions in the Arctic and regarding infrastructure investment and governance of port development. Marine development activities focusing on both trans-Arctic and destination shipping and resource extraction constitute active responses to climate changes that are increasing access to the north. Governance and regulation of this development requires international coordination and investment to reduce anticipated externalities from the development. The externalities presented by increases in shipping and resource extraction include consequences that may shrink the productivity of marine living resources through changing the ecosystem quality, as may happen with successful invasion by a new species. They also may increase the rate of change of broader climate changes underway. An example of this is how increased black carbon from increased vessel traffic will locally reduce ice cover, transforming its current uses, and will increase feedback into global climate warming through e.g. reduced reflection (the albedo effect).

Historically, development has preceded governance; the Arctic presents a rare and important opportunity to reverse this historically costly timing mismatch. The Arctic Council formed 20 years ago to provide a forum for negotiation amongst Arctic states and their indigenous peoples. The council provides a launching mechanism for this reversal of timing, so that governance structures aimed at societal well-being can shape economic development rather than be beholden to it. Table 1 lists the rotating chairmanships of the Council over the last 20 years alongside their highlighted strategic priorities.

Table 1 Arctic council chairmanship strategic priorities, 1996–2017

	Arctic council chairmanship	Strategic priorities/highlights
1996–1998 (Inauguration)	Canada	Arctic Council Establishment & Structure
		Sustainable Development & Environmental Protection
		Cooperation & Coordination among Arctic States
		Arctic Communities – Indigenous Peoples
		Dissemination of information & education
		(Arctic Council 1996)
1998–2000	USA	Sustainable Development
		Environmental Protection
		Education, Outreach & Coordination
		(Arctic Council Secretariat 1998)
2000–2002	Finland	International Cooperation
		Arctic Council's Structure
		Environmental Protection
		Sustainable Development
		Arctic Research & Education
		Economic & Social Development
		Indigenous Peoples & Regional Participation
		(Ministry for Foreign Affairs of Finland 2001)
2002–2004	Iceland	Human Development in the Arctic region
		Information Society, Human Resources, Research Collaborations-Arctic Education
		Cooperation & Dialogue with non-Arctic stakeholders
		(Ministry for Foreign Affairs of Iceland 2002)
2004–2006	Russian Federation	Circumpolar transport infrastructure development
		Sustainable management of resources – renewable sources of energy
		Climate Change & Environmental Protection
		Prevention & Management of Emergencies
		Arctic Communities – Indigenous Peoples
		(Ministry for Foreign Affairs of the Russian Federation 2004)
2006–2009	Norway	Integrated Resource Management, Climate Change, Emissions reductions & Removal of greenhouse gases
		Arctic Communities – Indigenous Peoples
		Arctic Council's Structure
		(Arctic Council 2008)
2009–2011	Kingdom of Denmark	Arctic Communities
		Climate Change, Biodiversity, Megatrends, Integrated Resource Management
		Cooperation within the Arctic Council
		(Danish Delegation-Arctic Council 2009)

(continued)

Table 1 (continued)

	Arctic council chairmanship	Strategic priorities/highlights
2011–2013	Sweden	Climate Change, Environmental Protection, Biodiversity
		Economic Development
		Arctic Communities – Indigenous Peoples (Sami)
		(Government Offices of Sweden – Ministry of Foreign Affairs 2011)
2013–2015	Canada	Arctic Communities – Indigenous Peoples
		Arctic Economic Council
		Climate Pollutants
		Marine-oil pollution
		(Global Affairs Canada 2015)
2015–2017	USA	Climate Change
		Arctic Ocean safety, Security and Stewardship
		Economic and living conditions of Arctic communities
		Raising public awareness
		(U.S. Department of State 2015)
2017–2019	Finland	Climate change and protection of the environment
		Economic & Sustainable Development
		Communication facilities, Education, Remote Arctic populations
		(Koivurova et al. 2017)

The Ottawa Declaration in 1996 marked the establishment of the Arctic Council, with Canada serving as its first chair until 1998. The Declaration sketches the structure of the Arctic Council and delineates its scopes in a broad context. The scopes center around cooperation amongst Arctic States and involvement of indigenous communities for addressing sustainable development and environmental protection, establishment and coordination of working groups and programs, and dissemination of information and education (Arctic Council 1996). This list is perhaps as noteworthy for what is not included as for what is. With a focus on peaceful coexistence, confrontational topics purposefully have been set aside (Humrich 2017).

Over the years, the focus of the scopes and goals of the Arctic Council have narrowed within the broader agenda. For example, broad fisheries management decisions have often been too controversial to include, in spite of their regional importance to sustainable development. Iceland's exclusion from the Arctic Five's discussions and subsequent decisions illustrates the tradeoffs: would Iceland have signed a moratorium on CAO fishing (Knecht 2017)? At the same time, however, profound qualitative changes have been witnessed in the way Arctic challenges are being addressed (Allison 2013). The working groups have undertaken increasing amounts of work and there has been a proliferation of new projects and initiatives.

Structural and operational changes over the years have also reshaped the Arctic Council and have partly altered some of its processes. Examples include the gradual reinforcement of the Indigenous Peoples Organizations as Permanent Participants, the establishment of the Arctic Council Secretariat in 2011 which became operational in 2013. Meanwhile, some of its participants have evolved and grown over time, such as for example the Inuit Circumpolar Council, whose role has been strengthened significantly since 1996 (Allison 2013). More importantly, over the course of the 20 past years, as the scientific understanding of the environmental challenges in the Arctic has been growing, the Arctic Council has been gaining more momentum with its objectives targeting specific problems such as black carbon and methane emissions, ocean acidification, invasive species, and so on. Furthermore, sustainable development issues have garnered considerable attention recently with living conditions and related challenges among indigenous communities having high prioritization on the agenda. The Arctic Council has become far more cognizant of the importance of the value of traditional knowledge, which is reflected in its continuous efforts to integrate it into assessment, planning and management, as well as in its endeavor to actively engage Indigenous Peoples in decision making processes.

The receding icy barriers have gradually allowed space for development of anthropogenic activities in the Arctic such as shipping, fisheries, tourism, and resource exploration. Many of these activities, whether local or international, are grounded in resource use and extraction. As such, they benefit from increased governance aimed at commons problems, industrial concentration, and related externalities (Kaiser et al. 2016). Furthermore, these activities can transform the productivity of ecosystems in the long run, so that missed or misguided governance opportunities in the present will have long term consequences for the future.

The Arctic states understand these challenges. The agendas of the Ministerial Meetings and the focus of the Working Groups of the Arctic Council indicate prompt responses to changes spurred by anthropogenic activity and prioritization of them accordingly, though all within the scope of peaceful interaction. Examples include safety and search-and-rescue strategies and oil pollution response planning, inter alia.

Besides its role as an intergovernmental forum, the Arctic Council also provides the platform for reaching legally binding agreements, though they are not able to make – or enforce – legally binding agreements directly. The first negotiated agreement was the "Agreement on Cooperation on Aeronautical and Maritime Search and Rescue in the Arctic" (Arctic Council 2011) which was signed at the Nuuk Ministerial meeting in 2011, and came into force in January 2013. The second legally binding agreement negotiated through the Arctic Council was the "Agreement on Cooperation on Marine Oil Pollution Preparedness and Response in the Arctic" (Arctic Council 2013) which was signed at the Kiruna Ministerial meeting in May 2013. At the Ministerial meeting in May 2017, state representatives signed a third agreement, which pertains to scientific cooperation amongst the Arctic states.

While the Arctic Council works to bring together the eight Arctic nations and permanent participants from indigenous Arctic communities to resolve governance issues requiring multilateral cooperation, the governance challenges the group faces are complex and cut across many different policy lines. Such challenges include non-legally binding policy. Furthermore, while other nation states may become, through an application process, observers of the council actions, engagement of the rest of the world directly with the Arctic Council is limited. The six working groups of the Arctic Council (ACAP, AMAP, CAFF, EPPR, PAME and SDWG) all have specific mandates that in principle should cover the realm of issues requiring multilateral decision-making. In some cases, however, certain concerns may fall into governance gaps outside of the Arctic Council entirely, between the working groups, or, in overlapping several groups, find themselves without dedicated resources or actionable governance plans.

Meanwhile, in view of the economic development in the region, the Arctic Council has also acted as a platform for shaping new institutional structures that promote Arctic economies and practices. One of the results of this process is the formation of the Arctic Economic Council, which is an independent organization facilitating business activities and economic development in the Arctic in a sustainable manner. This forum has no governance or enforcement capabilities either, however. Whether the separation of economic and other interests in the Arctic is a successful technique for achieving sustainable development remains to be seen. Currently, dampened resource prices for oil and gas and international political actions such as sanctions against Russia for its 2014 invasion of Ukraine have caused the Economic Council, and northern communities in general, to look for ways to diversify economic activity in the north. So far, the scale and scope of these activities appear limited in comparison to the potential riches of mineral and resource extraction, yet must meet these limitations if sustainable development in support of local communities is to succeed (Larsen and Fondahl 2015).

The way forward for the Arctic Council is likely to feature the development of more Task Forces as well as strengthening of the existing Working Groups. In this way, they can tackle emerging challenges driven both by man and nature, especially as more scientific knowledge becomes available. Over the years, the Arctic Council has continuously broadened its horizons, adapting its strategy to address the new challenges and prioritizing accordingly. Its umbrella strategy to bring together Arctic States and stakeholders and help forge common approaches to shared challenges across the circumpolar Arctic has existed since its establishment and has augmented over time. Given the recent successful legally binding instruments developed under the auspices of the Arctic Council, there is increasing discussion about whether there should be more focus on developing further such instruments of governance. While there are diverging viewpoints on whether it should move towards that direction, we expect that the Arctic Council will continue to reshape and adjust its strategy and scope to the new challenges, following the leads of scientific developments, as it has been doing up until now. Working together as sovereign Arctic Council Permanent Participants offers a complementary avenue of potential coordination in other global marine resource governance fora (United Nations, etc.) aside from the Arctic Council simply by virtue of ongoing communication to tackle current and future marine resource issues.

References

Allison, D. (2013). *Canada and the Arctic Council: An Agenda for regional leadership*. Report of the Standing Committee on Foreign Affairs and International Development, 41st Parliament, 1st Session.

Arctic Council. (1996). *Declaration on the establishment of the Arctic Council*. Ottawa. Retrieved from https://oaarchive.arctic- council.org/bitstream/handle/11374/85/EDOCS-1752-v2-ACMMCA00_Ottawa_1996_Founding_Declaration.PDF?sequence=5&isAllowed=y

Arctic Council. (2008). *Programme for the Norwegian Chairmanship of the Arctic Council 2006–2008*. Retrieved from http://library.arcticportal.org/319/1/AC_Programme_2006-2008.pdf

Arctic Council. (2011). *Agreement on cooperation on aeronautical and maritime search and rescue in the Arctic*. Retrieved from https://oaarchive.arctic-council.org/bitstream/handle/ 11374/531/EDOCS-1910-v1-ACMMDK07_Nuuk_2011_Arctic_SAR_Agreement_unsigned_EN.PDF?sequence=8&isAllowed=y

Arctic Council. (2013). *Agreement on Cooperation on Marine Oil Pollution Preparedness and Responsein the Arctic*. Retrieved from https://oaarchive.arctic-council.org/bitstream/handle/11374/529/EDOCS-2068-v1-ACMMSE08_KIRUNA_2013_agreement_on_oil_pollution_preparedness_and_response_signedAppendices_Original_130510.PDF?sequence=6&isAllowed=y

Arctic Council Secretariat. (1998). *Memo on U.S. Chairmanship Priorities, 1998–2000*. Washington, DC. Retrieved from https://oaarchive.arctic-council.org/bitstream/handle/11374/1890/EDOCS-4176-v1-1998-11-30_Memo_on_US_Chairmanship_priorities.pdf?sequence=1&isAllowed=y

Christiansen, J. S., Mecklenburg, C. W., & Karamushko, O. V. (2014). Arctic marine fishes and their fisheries in light of global change. *Global Change Biology, 20*(2), 352–359.

Danish Delegation-Arctic Council. (2009). The kingdom of Denmark. Chairmanship of the Arctic Council 2009–2011. In *The Danish Chairmanship Programme, presented by the Danish delegation to the Arctic Council at the Arctic Council Ministerial Meeting in Tromsø, Norway, April 29 2009*. Tromsø. Retrieved from https://oaarchive.arctic- council.org/bitstream/handle/11374/1565/ACMM06_Tromsoe_2009_Denmark_chairmanship_programme.pdf?sequence=1&isAllowed=y

Global Affairs Canada. (2015). Canada's Arctic Council Chairmanship. Retrieved from http://www.international.gc.ca/arctic-arctique/chairmanship-presidence.aspx?lang=eng

Government Offices of Sweden – Ministry of Foreign Affairs. (2011). Sweden's strategy for the Arctic region. Retrieved from https://openaid.se/wp-content/uploads/2014/04/Swedens-Strategy-for-the-Arctic-Region.pdf

Humrich, C. (2017). Coping with institutional challenges for arctic environmental governance. In: K. Keil & S. Knecht (Eds.), *Governing arctic change: Global perspectives* (pp. 81–99). Palgrave Macmillan UK.

Kaiser, B. A., Fernandez, L. M., & Vestergaard, N. (2016). The future of the marine Arctic: environmental and resource economic development issues. *The Polar Journal, 6*(1), 152–168.

Knecht, S. (2017). Exploring different levels of stakeholder activity in international institutions: Late bloomers, regular visitors, and overachievers in Arctic Council Working Groups. In K. Keil & S. Knecht (Eds.), *Governing Arctic change: Global perspectives* (pp. 163–186). Palgrave Macmillan UK.

Koivurova, T., Śmieszek, M., Stępień, A., Mikkola, H., Käpylä, J., & Kankaanpää, P. (2017). *Suomen puheenjohtajuus Arktisessa neuvostossa (2017–2019) muutoksen ja epävarmuuden aikakaudella*. Helsinki. Retrieved from http://www.fiia.fi/assets/Taustaraportti_250117PDF.pdf

Larsen, J. N., & Fondahl, G. (Eds.). (2015). *Arctic human development report: Regional processes and global linkages*. Nordic Council of Ministers.

Ministry for Foreign Affairs of Finland. (2001). *Program for the Finnish Chair of the Arctic Council 2000–2002*. Retrieved from http://www.arcticfinland.fi/loader.aspx?id=677109d0-e161-4e25-8ce7-91481f2860a8

Ministry for Foreign Affairs of Iceland. (2002). *Program for the Icelandic Chair of the Arctic Council 2002–2004*. Reykjavik. Retrieved from https://oaarchive.arctic-council.org/bitstream/handle/11374/1777/EDOCS-3652-v2-ACSAOFI04_Inari_2002 _14_Icelandic_Chairmanship_program_2002-2004.pdf?sequence=5&isAllowed=y

Ministry for Foreign Affairs of the Russian Federation. (2004). *Program of the Russian Federation chairmanship of the Arctic Council in 2004–2006*. Retrieved from https://oaarchive.arctic-council.org/bitstream/handle/11374/1766/EDOCS-3390-v1-ACMMIS04_ REYKJAVIK_2004_6_Russian_Chairmanship_Program.pdf?sequence=1&isAllowed=y

U.S. Department of State. (2015). *U.S. Chairmanship of the Arctic Council*. Retrieved from https://www.state.gov/e/oes/ocns/opa/arc/uschair/index.htm

Index

© Springer International Publishing AG 2018

N. Vestergaard et al. (eds.), *Arctic Marine Resource Governance and Development*, Springer Polar Sciences, https://doi.org/10.1007/978-3-319-67365-3

Printed in the United States
By Bookmasters